Crowd Management:
Risk, Security and Health

William O'Toole, Stephen Luke,
Travis Semmens, Jason Brown
and Andrew Tatrai

(G) Goodfellow Publishers Ltd

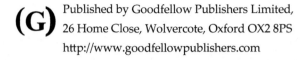

Published by Goodfellow Publishers Limited,
26 Home Close, Wolvercote, Oxford OX2 8PS
http://www.goodfellowpublishers.com

British Library Cataloguing in Publication Data: a catalogue
record for this title is available from the British Library.

Library of Congress Catalog Card Number: on file.

ISBN: 978-1-911396-88-8

The Events Management Theory and Methods Series

 Design and typesetting by P.K. McBride, www.macbride.org.uk

Cover design by Cylinder

Printed by Marston Book Services, www.marston.co.uk

Contents

List of Figures

Introduction to the Events Management Theory and Methods Series

Event management as a field of study and professional practice has its textbooks with plenty of models and advice, a body of knowledge (EMBOK), competency standards (MBECS) and professional associations with their codes of conduct. But to what extent is it truly an applied management field? In other words, where is the management theory in event management, how is it being used, and what are the practical applications?

Event tourism is a related field, one that is defined by the roles events play in tourism and economic development. The primary consideration has always been economic, although increasingly events and managed event portfolios meet more diverse goals for cities and countries. While the economic aspects have been well developed, especially economic impact assessment and forecasting, the application of management theory to event tourism has not received adequate attention.

In this book series we launch a process of examining the extent to which mainstream theory is being employed to develop event-specific theory, and to influence the practice of event management and event tourism. This is a very big task, as there are numerous possible theories, models and concepts, and virtually unlimited advice available on the management of firms, small and family businesses, government agencies and not-for-profits. Inevitably, we will have to be selective.

The starting point is theory. Scientific theory must both explain a phenomenon, and be able to predict what will happen. Experiments are the dominant form of classical theory development. But for management, predictive capabilities are usually lacking; it might be wiser to speak of theory in development, or theory fragments. It is often the process of theory development that marks research in management, including the testing of hypotheses and the formulation of propositions. Models, frameworks, concepts and sets of propositions are all part of this development.

The diagram illustrates this approach. All knowledge creation has potential application to management, as does theory from any discipline or field. The critical factor for this series is how the theory and related methods can be applied. In the core of this diagram are management and business theories which are the most directly pertinent, and they are often derived from foundation disciplines.

All the books in this series will be relatively short, and similarly structured. They are designed to be used by teachers who need theoretical foundations and case studies for their classes, by students in need of reference works, by professionals wanting increased understanding alongside practical methods, and by agencies or associations that want their members and stakeholders to have access to a library of valuable resources. The nature of the series is that as it grows, components can be assembled by request. That is, users can order a book or collection of chapters to exactly suit their needs.

All the books will introduce the theory, show how it is being used in the events sector through a literature review, incorporate examples and case studies written by researchers and/or practitioners, and contain methods that can be used effectively in the real world.

Preamble

When I began organising events, we would simply book a hall, book a band and sell tickets. It was that simple. Not any more. As the events sector has grown, there have been mighty phase changes. From the simple 'gig', the concerts started to attract people from outside the town. Tourist spending made a change to the local economy. As it grew, governments realised that the tax revenues, social and economic activity were vote winners. Concerts became five day festivals. Then they toured the world. Small groups of people watching their favourite band became tens of thousands and then hundreds of thousands of people. In the sports events, the local match became part of the global competition that culminated in the Rugby World Cups and the Olympics. Conferences and exhibitions went from local trade fairs to enormous Abu Dhabi IDEX or Hanover CeBit-like gatherings.

From all this emerged the science of event management. No longer 'ad hoc' and just responding to the issues as they arose, event management is now a plethora of plans submitted to the stakeholders for prior approval. The famed work breakdown structure of the event planning produced highly specialized areas that require specific expertise and experience. It was the dialectic of quantitative growth producing qualitative change. As the crowds grew and the events became more visible and economically important, the spin-off was the formation of new companies and experts specialising in just one aspect of event management. It is hard to comprehend how quickly this has occurred. Over the last ten years, one website in Dubai (SourceME) grew to host 3,500 companies that service the events sector. In 2019 with billions of dollars, Saudi Arabia created 11 major festivals spread over the whole year around the country (www. saudiseasons.sa).

The result of this rapid growth is a significant lag in information from the practitioners to researchers, academia and students. They are just too busy 'doing it'. Fortunately, working with the Australian Institute of Disaster Resilience allowed me the time to discover the common systems used in each of these areas. The reader must understand that the information contained in each of the chapters is from these people. I merely recognized the common pattern and issues.

I am privileged to know many of the experts working in the events sector. I have asked four of these experts to write the sections of our book. They are people who, on a daily basis, live their areas of expertise. They, as Nassim Taleb writes, have 'skin in the game'. Their words have consequence.

I would like to thank the staff of the Australia Institute for Disaster Resilience and their excellent series of handbooks that are quoted many times around the world.

I would also like to thank all those who work on events so that we, the audience, may be able to enjoy the pleasure of attending with our friends, colleagues and family.

Introduction

There are crowds at the beach, while I am waiting for a train, there are lunch crowds, crowds at concerts, you may walk down the street turn a corner into a protest crowd, there may be a new phone launched and suddenly there is a crowd. You can't escape it in a city. After the city you then go to a concert in the country, a festival, a big BBQ or a football match. Crowds mean proximity, queues or lines, they mean delay, frustrations, they also mean celebration, socialisation and fun.

They can also mean disaster.

In a crowd you can lose the freedom of choice and movement, your risk of catching infectious diseases increases and you can be a target for theft or a hostile act.

The aim of the book is to present the different perspectives on crowds. Although we will focus on events, such as public festivals, sports and concerts, this information and the system is applicable to mass gatherings anywhere in the world. It is to help the reader understand the different aspects of predicting, managing and evaluating a crowd. We have chosen four aspects of crowds. We call them *domains* as they have their own history, terminology, knowledge, experience, competency and processes.

The first two chapters describe the framework common to the four domains.

The perspective we take on crowds is from the Complexity theory. A crowd is complex as each individual is a free agent and, in a crowd, responds to the people near them. Instantaneously, their attitudes and behaviours affect that individual. Hence it is the relationships between the people that create crowd behaviour. This is beyond complicated, because the causes and effects are immersed in multiple behaviours and attitudes that are moving and developing. One aspect of the complexity is called *emergent behaviour*. It is behaviour that cannot be predicted from each individual. It is as though the crowd itself has its own life. In a complex situation it is impossible to predict exactly what will happen all the time as a very small change in a crowd may be amplified throughout the whole crowd. Take for example a car backfiring near a crowd. This could be ignored by some crowds or at other times may cause a panic and the people to rush away from the origin of the sound. The rush could then cause other people to panic and, like the famous experiment with mouse traps and ping pong balls, produce a calamity.

"About 1,500 people were injured in the Italian city of Turin after a firecracker provoked a stampede on Saturday night, police say. Thousands of football fans were watching a live relay of Juventus' Champions League final against Real Madrid in Cardiff when a bang was heard and rumours of an explosion spread"
https://www.bbc.com/news/world-europe-40147813, 2017

"A rush-hour stampede at a railway station in Mumbai has left at least 22 people dead and dozens more injured. Indian police said it was triggered by a rumour that a pedestrian bridge was collapsing, sparking chaos as passengers surged forwards to flee"
https://www.telegraph.co.uk/news/2017/09/29/mumbai-station-stampede-kills-least-15-amid-rumour-bridge-collapsing/, Sept 2017

"If you take anything organic and you try to control its variability, you'll end up with less variability than you started with but the system would become more fragile" Nassim Taleb (2012)

In event management theory we now stand at the junction between rigid planning and flexibility. This is obvious when it comes to crowd management. Planning is vital and there are many articles and books on this. But it is not enough. Because so many plans are now required by governments, it can easily be a 'tick box' operation. Last year's plan is used with the dates changed. After many disasters we now realise that the plan is only part of the story. The plan must be implemented, changed, adapted to changing circumstance and even ignored if that is required.

Who is this book for?

♦ Event professionals who need an overview of crowd management.

♦ Students of event management who need to prepare the way for a career.

♦ Government departments involved in crowds, such as Tourism and Transport.

♦ Trainers and teachers in the field of events, to develop the curriculum.

♦ Academics who need the information and theory from those at the front line of crowd management and events.

♦ Readers interested in the application of complexity theory to their everyday world.

1 The Management Framework

Introduction

The management framework is introduced. This comprises the common high level processes found in the four areas or domains on crowd management. The terminology and concepts described are: state, complexity, emergence, input/process/output, factor analysis and phase change. Using this framework assists the reader to understand the four domains described in the four sections and integrate the knowledge, theory, terminology and examples into a management system.

Complex

The writing illustrates a systems theory of event management. In this case it is process mapping under risk and complexity. The text describes the processes involved in each domain of security, medical, crowd management: inputs, processes, outputs. The processes are optimised by decision theory under uncertainty. Put in simpler terms, this means that the way to do things is made more efficient by making good decisions. But, often in events and crowds, you can never be certain. Decisions are mostly approximate or the best estimate. Another name for this is 'local optimisation'. One result of the complexity, such as found in crowds and events, is the emergence of states that cannot be predicted. An example is a children's party. Anyone who has been involved knows how quickly it can get out of hand. One minute everyone is sitting chatting and eating and then, in a flash, they are up running around. By that we mean an emergent state. It was impossible to predict exactly. You just know something will happen if you take your eyes off the children. You need to be able to react. The more experience you have the better you are at knowing how to minimise the chaos.

Example 1.1: Children's concert

In Bahrain there was a concert for children at night, when it is cool, in a very large tent. Over a thousand children were attending. Suddenly the lights went out due to a power failure. What would you do? How would you manage this situation. It is no good saying you would have 'backup' – you didn't. What exactly is the problem? How long do you have to solve it?

This experience, learning from mistakes, keeping the mistakes small, leaning from others in similar circumstances, becomes the basis of resilience. The domains of crowd management, security and health are all 'costs' to the event. Hence risks must be managed and the solutions optimised to minimise catastrophic or 'long tail' consequences.

Optimisation means finding the best solution to a problem. At the operations level, a person may not have all the information and they may be unsure of the result of any decision. But a decision must be made. Hence they need to make the best decision at the time. An example is a decision that must be made in a crowded place to close the entrance. Limited information is available, but the decision must be made.

A process is a series of tasks or actions. Event management as outlined in William O'Toole's *Events Feasibility and Development*, comprises a group of processes. Many of these process are found in project management. The *Project Management Body of Knowledge* describes these very well. The processes are intertwined. Although described as 'laneways' they, in fact, interact constantly and these interactions contribute to the complexity. For example the process of sponsorship interacts with marketing, which interacts with staging. To make it more confusing, these interactions are not one way. The staging process will interact with the sponsorship. Hence the management of this aspect of events is an excellent example of the complexity management theory.

Characteristics of complexity management

At a specific time the state of a crowd can be described by a number of characteristics. These can change and hence they can be called *variables*. One can describe a crowd as happy, quiet, discontented, for example. Other variables could be *size* such as the number of people or the area of the event site and demographic characteristics. Listing the characteristics, as shown below, gives what we call the *state*.

A 'state' is the description of the variables or factors at a specific time – a term used in Markov Chains adapted to events. It can be expressed as a collection or matrix of variables, fixed in time. An example is describing a crowd with three variables: *mood*, *density* and *flow* for the purpose of crowd control decisions.

In technical terminology, a complex situation/system is characterised by:

♦   Innumerable linkage (affects) of tasks, interdependence.

♦   Inability to forecast using linear analysis.

♦   Ability of the situation to quickly amplify and dramatically change.

♦   Risks that are rare yet with catastrophic consequences (long tail risk).

♦   Uncertainty and the inability to assess the state transitions using statistics to assess the probability matrix.

♦   Unique state (expressed as a matrix): in events suppliers, site, theme, demographics, time-based situation.

♦   Emergent behaviour – i.e. new state factors, arising from, for example, the increase in the scale of the factors.

The management of crowds such as is found in events and festivals is an example of the application of the management of complexity. Traditional planning is initially used, however a crowd is a complex situation and a fixed plan works best when the crowd is compliant. More detail is found on this in Andrew Tartai's application of the Snowden's Cynefin Model to crowd management in Chapter 5. The event management team must take into account the triggers, the complexity, the emergent behaviour and the catastrophic risks. For example hostile attacks on crowds, such as terrorism, demonstrates that planning is necessary but not sufficient in the management of crowds. Every event management team now needs to understand complexity, decisions under uncertainty and long tail risks.

Framework

The metaphor of a framework means an underlying system or skeleton common to these four sections of the textbook. The four following perspectives came from very different disciplines. They all converged at an event. The events sector is relatively new and it draws on theory and experience from a wide variety of disciplines such as theatre production, marketing, engineering, film production, project management, logistics and much more. This is illustrated by numerous themed chapters of Professor Don Getz's textbook, *Event Studies*. The richness of the events sector is plainly shown by the scope of his book. In our case, the disciplines are crowd management, risk, medical and security. They are disparate but related. They have their own terminology, authority structure and ways of working. But they must work together in front of a crowd. Surprisingly, for each of these domains, behind the scene, is a similar structure. They are each trying to pre-empt and solve problems that arise from a mass of people. The problems and their solutions define the underlying structure. This book extracts that framework. But, why do it?

1   It enables each of the four domains to be understood by the reader and the information to be learned, used and placed in a common system.

2   The framework assists in developing and maintaining quality teaching and training programs.

3   It supports data collection and future research.

4   It enables event practitioners and suppliers to work and consult with companies and experts in these fields.

Input/process/output

For the purpose of establishing the framework that will enable clear teaching and examples, I will use an engineering version of the definition of complexity. For the clarity of this model, people are regarded as elements in a complex system. Although motivation and other intangible aspects of people are vital to crowds and risk, I will start with their behaviour as individuals and the behaviours of the crowd. This may seem I am leaving out the all-important human part, but in order to

establish a clear framework for the four domains of crowds it is necessary. The input/process/output/ (IPO) model creates a common structure
to the four very different sections. In each of the sections the reader will
recognise:

♦ Inputs

♦ Processes

♦ Outputs

♦ Tools and techniques

♦ Complexity and emergence

♦ Factor analysis

♦ Decision making in a dynamic time-critical environment

All of these are explained in detail in the first two chapters. They
can then be used to analyse and synthesize the information in the four
domains. Each domain has enormous amount of literature written about
it, with journals, websites, companies, government departments, experts
and consultants. This textbook is the first to draw them all together
under the one framework. The framework is open to new information
and can expand as more data is created.

The framework facilitates teaching and training as it provides a tem-
plate or series of headings that can transferred into lecture and tutorial
topics. It allows the myriad of facts, vignettes, and experiences crossing
all these areas to fit together.

Borrowing also from the input/process/output model, the chapters
will emphasise the following:

Environment

The surrounds in which the crowd is found; physical, social, economic
and political. Below is an example of the affect the environment can have
on crowd management.

The most obvious recent environment change affecting events and
crowds has been the sudden increase in terrorist and other attacks at
events. This is a long tail risk, meaning that it is currently unpredictable
and the results are disastrous. The response to the environment has
been varied around the world. Events, such as community festivals, in

areas that have never had an attack of any kind have to prepare with extra security. Bollards, such as cement blocks, and using trucks to block streets, can be found in the most remote towns with no history of terrorism. There is secondary risk as illustrated in the crowd stampede in 2019 at the Rolling Loud in Miami. The cause is unknown but the fear was that there was an 'active shooter' in the crowd. It spread quickly and resulted in panic and the crowd racing for exits and protection. In this sense it is not the actual physical violence that is the issue; it is the crowd's perception.

Inputs

The information, resources, plans needed, list of event characteristics and their effect on the risks (factor analysis), stakeholder consultation results, crowd characteristics, recent news reports are all inputs. There are many thousands of factors that influence the people in a crowd; the role of a model is to, first, put this into an order of importance. Hence the different domains will order these according to the effect that is felt in the domain.

The attendees of an event at the food stalls or vendor area may be quite pleased that there is only a small crowd. However a near empty field in front of a stage will feel disappointing to the audience. At concerts the crowd and their enthusiasm help to build the 'vibe' or atmosphere.

An excellent example of inputs for a specific domain is found in *Public Health for Mass Gatherings: Key Considerations*, published by the World Health Organization. In the detailed table, one entry is 'International' as a description of the crowd and with a list of health risks such as:

♦ Risk of importation/exportation of disease;

♦ Risk of delayed access to healthcare due to unfamiliarity with healthcare system;

♦ Risk of delayed detection of pathogens by inexperienced healthcare system;

♦ Risk of environmental risks for those not acclimatized such as heat or cold, altitude, pollution;

♦ Communicable disease for unvaccinated or vulnerable travellers to endemic pathogens and parasites;

◆   Unknown immunity of participants.

From World Health Organization 2015, Page 19 Table 1: Examples of MG event assessment characteristics

Processes

These are time-ordered tasks performed to produce a result. Some of the processes are common to all the sections, such as risk management, others are specific to the section or domain, such as the presentation rate in the medical domain that determines the type and amount of resources needed. Some processes may be running at the same time, whereas others may need the output of another process before they can start.

Process mapping is a common tool used to develop a model of management. Once the processes have been discovered and described, it is relatively easy to convert them into software programs, algorithms and apps.

Tools and techniques

These tend to be specific to the domain. In the Health section, for example, the reader will find numerous specialist techniques such as the chain of survival. Each of the four sections has evolved separately and hence has terminology as well as tools and techniques that are different. Overseeing all this is the risk management. At a finer detail are the heuristics, skills learned on the job to help complete a specific task.

An example of heuristics is the seemingly inexplicable sudden increase in crowd density and congestion at certain places in the event site. A ferry arrives at a wharf quite a distance from the event. The arrival is not obvious to security at the event on the ground. All they see is the pulse in numbers, sudden slowing down of the flow which can be felt at certain pinch points in the crowd. With experience the crowd management at events will understand and prepare for these situations.

Outputs

Outputs are the results of a process. Outputs can also be termed 'deliverables' in project management. A risk management plan is an example of an output of the risk management process. One output of the site design process is the site map. Most of the outputs during the planning phase of an event are, unsurprisingly, the plans, such as marketing plan, budget, schedule and stakeholder management plan. The problem can be that the plans are too often mistaken for the action that the plans call for. Planning becomes the aim and not the management of the event. No matter what plans are produced, they still have to be implemented, changed, constantly assessed and improved. Also this is under the threat of the deadline. Unfortunately the comment "more planning is needed" stops conversation and is seen as the solution to all problems. In events and crowds planning is only part of the solution.

The linear nature of the input/process/output model works when the situation is stable. For example putting more funds into publicity will increase the number of tickets sold, which will increase the crowd size at an event. Opening the right barriers or gates will assist the movement of the crowd around the site. In most cases, event management is about this straightforward system. Such a linear system does not consider the deadline. Everything in the management of an event has the element of time. The closer you get to the day of the event, the more the tasks become time-critical.

An important part of the four domains and the framework is time. Event management decisions always have to consider the deadline. This is often left out of management models but it is vital in events. The number of stakeholders, including the audience, creates an environment where the schedule drives the event. Unlike other projects where major delays mean shifting the completion date and penalties, an event is rarely delayed in the same way. It could mean cancellation. Cancellation of an event means that the value of the event is unrealised. In colloquial terms "all the work was for nothing". The result can be: tickets refunded, contracts cancelled, loss of reputation, costs still have to be met and possible financial ruin.

Factor analysis

A factor is a characteristic of the crowd or the environment that significantly contributes to the risks described in the following chapters. A windy day will have an effect on an outdoor crowd and the ability to control the crowd. A recent hostile attack and its publicity will influence the behaviour of a crowd and the security management. Demographic analysis may indicate there will be a significant problem with drugs at an event. The table referred to from the *Public Health for Mass Gatherings: Key Considerations* manual lists the initial factors that will influence many aspects of the health plan for an event.

Example 1.2: Spread of measles

Festival visitors urged to get measles jab

30 May 2018 09:32

👍 Like | Share | 4 people like this. Sign Up to see what your friends like

Visitors of music festivals have been warned to ensure they have had their measles jab amid concerns of outbreaks of the highly infectious disease around the country.

According to Public Health England (PHE), younger people are more at risk of catching the disease due to "close-mixing environments such as festivals ". It also said that it would be issuing alerts nationally through the summer.

The government health body said that festivals make the "ideal opportunity" for visitors to catch and spread the disease.

Essential checks

Among the areas in the UK with a higher-than-normal number of measles cases in recent months is Coventry, which is home to the BBC's The Biggest Weekend two-day festivity.

Measles is "a highly infectious viral illness" that can be fatal

---

Example 1.2 *Spread of measles* illustrates the almost daily health risk stories about crowds at events. A disease that was widely regarded as conquered is now spread though the advent of crowds at festivals.

Factor analysis concerns tracing the risks back to their causes. In mathematics it is a statistical tool and it used extensively in the public health area. In the other domains it is not developed to this level due to the lack of data.

A useful analytical tool for this is the Isikawa or cause-and-effect diagram. A risk is described, such as a drug death at a concert, and the various possible causes are listed. The history of this type of risk is researched to establish the causes of past drug deaths. These are graphically shown and their contribution to the risk is estimated. From all the Isikawa diagrams made for risks, the common factors that contribute to the risks can be found. Root cause analysis finds the single determining cause of a risk. That means if that cause is eliminated, the risk will disappear. For the drug deaths at music festivals, is the root cause the ability to smuggle in the drugs? If we eliminated the smuggling would this stop the practice? Once again we come up against the complexity of the issue. The cause effect diagram would show numerous contributing factors.

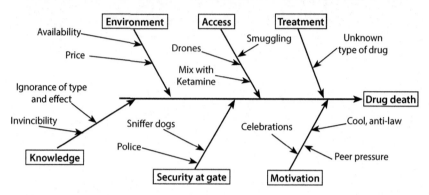

Figure 1.1: Cause effect diagram for drug deaths

A comprehensive list of event characterises as they affect health and well being of a crowd is given by the World Health Organisation see Figure 1.2.

□ Type of event
□ Activity Level
□ Duration
□ Occurrence
□ Season
□ Participant; Origins, (Local or International) Health status and density
□ Venue Characteristics (out door Indoor..)
□ Alcohol
□ Catering
□ Likely Drug use
□ Current medical services
□ Hygiene and Sanitation Services.

Figure 1.2: WHO Event Characteristics

There are many other characteristics that are not seen as important enough to make it into the WHO list. Motivation, for example, as described by Hutton, Ranse and Brendan (2018) is one. Terrorism target characteristic is another. This illustrates that each domains sees a crowd slightly differently. They look at the risks and work backwards to what are the characteristics that contribute to the risk in their field of expertise. So a crowd can have many characteristics but which of these become factors in a crisis, for example, is complex. The priorities of these factors will vary over times and then if there was a change, the influence of the each of the factors will change.

The WHO list in Figure 1.2 can be compared to the characteristics used by security assessment such as:

♦ Location

♦ Physical vulnerability to attack

♦ Social, politics and cultural importance of the event or venue

♦ Predictability of the event.

These are related but quite different. In the methodology of factor analysis, each of these characteristics would be measured from a number of similar events and their contribution or power to influence the health or security of the event would be estimated. In a standard manufacturing and service industry this factor analysis, makes a lot of sense. Just from the list above it can be readily seen that the nature of crowds is more fluid and temporary. Also the factors are not independent of each other; for example, audience profile and artist profile are related. A symphony orchestra will have a different audience to a gansta rapper. A point that may seem obvious, but has been missed quite few times.

For example, the *UK 1991 Event Safety Guide* gives the following advice:

Ensure that you have considered the following factors:

♦ *proposed occupant capacity*

♦ *artist profile*

♦ *audience profile*

♦ *duration and timing of event*

♦ *venue evaluation*

♦ *whether alcohol is on sale*

♦   *whether the audience is standing, seated or a mixture of both*

♦   *the movement of the audience between the entertainment and/or facilities*

♦   *artistic nature of the event, single stage, multiple-arena complex, etc.*

The above information can then be used to determine the provisions and facilities needed within the site, for example stages, tents, barriers, toilets, first aid, concessions, exits, entrances, hospitality area, sight lines, power, water, sewerage, gas, delay towers, perimeter fencing, backstage requirements, viewing platforms and waste disposal requirements. Once all the information is collated, detailed site design can begin. (Health and Safety Executive, 1999)

Phase change

Using a simple mathematical model a crowd can be described as being in a *'state'* at any one time. This is an abstract concept but corresponds approximately to the real world. There will be a number (quite a few actually) of factors that create this state, such as the weather and motivation. These can be regarded as variables and, if their effect is established, they are referred to as factors. Others may include density, satisfaction and disruptive elements. All of which are called *state variables*, i.e. variables that define the state of the crowd at a fixed time.

The variables used to describe the state will depend on the risks that are paramount to the observer. Hence a calm quiet seated crowd could be regarded by security people as a possible target, whereas to the first aid professionals it will be a low risk situation.

Moving from one state to another, such as when the concert has finished on stage, could be regarded as a *state transition*. Or the actual movement or exiting the venue could be seen as a *dynamic state*. In either case the state variables are changed. If *flow* and *density* are used as the state variables, then these may be a constant. If these are less than a certain value, the crowd controllers will only be monitoring. If there is a sudden change, then they will trigger new modifying actions by the crowd controllers. These variables are not independent of each other and therefore once they reach certain values there is the risk of a change. The crowd, although always complex, can change from a predictable and calm state to being dynamic and impulsive.

Although these variables are described as separate for the sake of the framework, they affect each other. This is common in a system. For example, the weather variable will affect many of the others. The flow variable will affect density and the other way around. It is impossible to precisely describe the effect of all the variables. Hence we must look on the system, the crowd as a whole. We can watch how the variables affect each other and describe the macro results. So for example as density diminished the flow can increase. Then at a certain point the flow decreases as the density decreased, i.e. there are fewer people passing a certain point as they are more spread out.

As the flow continues the density of the crowd decreases. Now these are just two very observable variables. There are a myriad of variables and hence it is complicated. To add to the complication, each person is a free agent. That may sound obvious, but it is vital to understand that this is not an engineering problem. People have free will and they will make up their own mind, possibly change it rapidly and act accordingly. Although this is an intangible variable, we know it can be manipulated. The people's trust in the authorities such as the security company is a good example. If some people in the crowd do not trust the authorities, they will not follow any suggestions. Hence one aspect of crowd management disappears. Communication works when people trust the message to be true and accurate.

Conclusion

As the events sector grows around the world and information is collated from thousands of festivals, concerts and other crowded places, patterns are emerging. This chapter introduced a framework to structure all the information being collected concerning crowds. It is the input/process/output (IPO) model combined with complexity theory. IPO is a well known framework and allows for effective planning. It concerns looking at the various causes and their effect, and planning to mitigate any risk. However the plan is not enough as a crowd is complex. It can go from behaving as a single placid group to sudden panic and chaos. Hence the new science of complexity needs to be applied. This type of situation needs competent staff to be aware of and recognise impending issues. The next chapter describes and examines the theory of the situation on the ground at the event.

Example 1.3: Measuring crowd numbers

How many people at the event?

Clarify the request

e.g. What will you use this information for?
What decision does this data inform? Risk? ....

Most
← important
step

Stakeholder	Accuracy
Client	
Security	
Insurance	
Marketing	
Supplier	
Others..	

factor	accuracy
Density	
Tickets	
Clickers	
Water sold	
Past data	
Others..	

compare

Maths tools: Kalman Filter, Bayesian Networks...

ANSWER	accuracy

The question could be for a small community festival or a mega event. Is it: how many people came to the event or how many people are at the event at a certain time or over a time period?

Measuring anything in the real world is estimating. There is always a tolerance, a plus or minus, or leeway with measurement. Most people assume anything with a number is accurate. So the first step it to realize this is estimating – not counting.

Next step is to understand that asking for a number is actually part of making a decision. That number is going to be used to make a management decision. What is the decision? What will the estimate be used for? The media are generally happy with a plus or minus accuracy of 30% or more. Security, insurance companies and police want far more accurate estimate .

Hence, what is the level of accuracy required by this stakeholder and decision to be made?

All of this is illustrated in the process in the diagram above.

Each stakeholder may want a different level of accuracy. However they may not be aware of this.

Next step is to decide on the methods of measurement: tickets, invitations, drone aerial readings of density, amount of garbage collected, clickers, CCTV, cars in the car park, past measurements correlated with these other measurements, number of water bottles sold, and much more. Each method of measurement has its own level of accuracy. Some measurements, such as the number of water bottles sold, is a proxy measure of the crowd numbers. It is indirect. However some festival organizers claim it is the most accurate they need for their event decisions. By combining these estimations and their levels of accuracy or uncertainty, a best estimate can be found. If it is required (e.g. the event has the budget to do it and the event risk profile needs it) certain mathematical techniques can be used. They come from statistics, probability, such as Bayesian and now used in robotics, Kalman filter. These help to combine all the estimates to gain a best practice and accountable final figure.

BUT it is still an estimate

2 Decision Making on the Day

"What matters most is not generating the best possible plan but achieving the best possible result "

United States Marine Corp, quoted in Riley, 2014, p. 40

Overview

Chapter 2 concerns issues on the day of the event and within the crowd. In a complex situation effective decisions must be made before small risks become disasters. In this chapter we explore the numerous techniques common to the four sections. These are heuristics, situational awareness, appropriate response, proportionality, local optimization, triggers and resilience. The domains may have different names for these and they will describe them from their perspective in the proceeding chapters.

Introduction

A common thread to each of the following chapters is the need to make decisions during the event or while there is a crowd. At some time on the event site, the person responsible may not have the luxury to weigh up all the options, assess the probabilities and take time to make a decision. The time needed for this can increase both the likelihood and the consequence of the risk. Not making a decision and lack of action may make the situation far worse. The techniques listed in this section will assist the decision maker at that time. They have a long history in similar environments such as policing, military and emergencies. These are complex, where time is important, information is limited and there is a potential for chaos. The following concepts are found throughout the four domains. They will be described slightly differently, although their aim is to minimise risk using the tools available. The reader should be able to list them when studying the specialist literature and website and recognise them when working in the field.

Heuristics

"Heuristics are simplified rules of thumb that make things simple and easy to implement. But their main advantage is that the user knows that they are not perfect, just expedient, and is therefore less fooled by their powers. " Nassim Taleb, *Antifragile*.

Heuristics is the knowledge and skill that is learned on the job. It tends to be localised and difficult to teach outside its area of immediate application. Tips and tricks is another term for it. Often it is so specialised that it cannot be described without actually performing the action. Most manual tasks are very involved and complex and yet a person with experience can do them with ease. Often they can't describe the techniques they are using.

A personalised checklist is an example of a heuristic. The new term for this everyday practicality is life-hacks. They tend to be left out of any theory and yet can be the difference between success and failure. The complex situation is an example of an environment where the heuristic fix of a minor problem can prevent disasters. One of the reason heuristics is not recognised as important is that if a problem does not happen, then it is a assumed it was never going to happen. The proverb "for want of a nail the war was lost" is an example. If the war was won, no one would say it was because the horse was shod correctly.

The overwhelming risks, often called *operational risks*, are found on the day of the event. Although they may be traced back to the project planning, the on-the-day staff will have to deal with them. To illustrate this point consider a simple electric cable or extension cord. The work health and safety standard is to tape the cord down or to cover it. But there are many times when this is not done for whatever reason. In some workplaces, such as an open space with one or two people, the temporary cable may not cause an accident. However in a crowd or on a stage at an event, a cable on the floor is a major risk. It is almost certain that someone will trip over it. That person will probably fall on other people and a small risk is quickly amplified. Only experience at events will pick up these types of details.

An interesting corollary to this is that by taping down or covering the cords, people do not look for cords. There is an argument that by taking these types of actions we are increasing the effect of the risk. We are transferring the risk into a long tail risk, i.e. from a minor common risk to a rare but disastrous risk. John Adams in his fascinating work on *Risk* (1995) explores his concept of the risk thermostat.

A further example is the lighting at events and crowds. Coloured lights can make certain colours disappear. An object such as a blue box on a stage under blue light will not be seen clearly and can become a hazard. The stage is often a place of a lot of activity, particularly during the setup time. Coloured lights can be dangerous at that time. These tips and tricks are picked up on the job.

One of the problems with describing heuristics is that they often seem obvious after they are explained. But in the time-constrained complex environment of an event, it may not be so obvious. These tips and tricks are highly valuable in these situations because there is often a choice of actions and the consequences can be dire.

Situational awareness

Situational awareness is found in the chapters on health and security. During an emergency some people will become too highly focussed. People report tunnel vision where everything outside of the one spot, the source of the trouble, just disappears. The same effect happens with people's perception of sound. When people are intensely focussed on one task they will not hear other things. This can be a major problem when there is an incident or emergency. Steven A. Adelman description of the 10-80-10 situation during emergencies is an example.

The 10-80-10 Rule

The "10-80-10 Rule" is the name of an observation, based on survivor accounts dating as far back as the 1911 Triangle Shirtwaist Factory fire and as recent as the latest active shooter incident, about the way people tend to behave in emergencies. Visually, it's just an ordinary bell-shaped curve.

On the left side of the curve, 10% of people will perceive an emergency quickly and accurately and then respond decisively to save themselves and others. Psychologically, these leaders suffer from less confirmation bias or normalcy bias than their peers, often because they have had some analogous life experience or training in the military or public safety. There are few of these people in most crowds, but they tend to self-identify as leaders.

This is important because the next group, roughly 80% of most crowds, are followers. They are relatively slow to correctly identify the threat – survivors of active shooter incidents, for example, routinely think they hear fireworks or another innocuous daily sound rather than gunfire – and then slow to take action. A common response during the September 11, 2001 attacks on the World Trade towers was 'milling', in which people knew the situation called for some action, but they focused on gathering even unimportant personal effects or shutting down their computers rather than running to safety. The presence of a leader exerting authority has been demonstrated to break followers out of their perfectly normal state of incomprehension and help herd them away from danger.

Finally, the last portion of the curve, the remaining 10%, are the few that tend to be shown on TV running and screaming. Historical accounts show unequivocally that panic is actually an infrequent occurrence even in the worst emergencies, for reasons that are thoroughly explained in the literature of disaster psychology.

It is important for event organizers and crowd managers to understand this 10-80-10 division in emergency responses. They can use this information to plan accordingly by identifying potential leaders by their experience, training, and temperament, and then providing those potential leaders with the tools and authority to help guide everyone else to safety.

Steven A. Adelman, Adelman Law Group, PLLC, Vice President, Event Safety Alliance, 8776 E. Shea Blvd., Suite 106-510, Scottsdale, AZ 85260, PH: 480-209-2426, Email: sadelman@adelmanlawgroup.com, Web: http://www.adelmanlawgroup.com/

Situation awareness is just the opposite. It is to be aware of all the things that surround the actions and to be able to find solutions and to take command if need be. Situational awareness is more than being aware during emergencies. The event staff should be aware at all times on the site. It is too easy to be lulled into the 'spirit of the event'. This sometimes conflicts with the needs of marketing, where the staff have to be part of the entertainment or event experience. Situational awareness means the person:

♦   Is alive to the immediate situation around them

♦   Can think clearly and assess any changes

♦   Understands the surrounding environment

♦   Can make decisions based on all this information.

Often a staff member or subcontractor in a crowd needs to make the best decision possible, using the information available within a critical time period.

Kime is a Japanese term in martial arts meaning to be aware, committed and focussed, a relaxed readiness. When the opponent looks defeated, it does not mean they are defeated. This concept is useful in crowd management as a problem may seem to be solved with a result that the staff will lose focus on the incident. In fact the problem may not be solved. "It's not over till it's over" is the colloquial term.

Spontaneous volunteers

In many disasters there are people who will immediately volunteer to help. This is common in natural disasters such as fires and floods. They can make a difference by providing skills and support. The event team must be aware of this and understand how to deal with these people. In a chaotic situation where quick action is vital, there may not be time for top down orders or command and control. Here the skills and knowledge of some members of the crowd can save lives. The event team needs to know how to handle being overwhelmed by the number of volunteers, how to use their skills and keep them informed.

During the stage collapse in Indianapolis State fair concert, the video clearly shows the actions of members of the crowd. One can see the panic and hear the people screaming in the stands. Then closer to the collapsed stage, the people can be seen physically lifting the structure so that trapped audience members can be pulled out. This is a crystal clear visual of spontaneous volunteering in action.

This concept is similar to the USA Department of Homeland Security's *see something say something campaign*. In this case the security team are using the collective 'eyes' of the crowd to assist identifying unusual objects or suspicious activity on site.

Appropriate response

Linked to heuristics and situational awareness is the concept of appropriate response. This is to ensure that the response to any risk is not overdone. The level of the response must be determined by the size of the risk.

To crack a nut with a sledge hammer, not only is unnecessary force but it also destroys the kernel. The term appropriate in part means that the solution to the problem or the risk minimisation does not waste resources or cause more problems. It comes from approach to risk in project management, where the response should be within the time-cost-scope boundaries. The response to the risks discovered on-site with a crowd will be time-dependant. A small issue can quickly become a major problem and therefore an appropriate response must take into account the time needed to address the issue. Combine this with the delay that is a result of crowded conditions, and the response has to be well thought out very quickly.

The qualities of an appropriate response are:

♦ Timely – some risks require immediate action.

♦ Correct amount of resources

♦ Manageable secondary or residual risk

Proportionality

The principle of proportionality is common in the legal sphere, such as sentencing, and is now used in the planning for the treatment of risk. It means that the proposed prevention of the risk should not be any more than is necessary. It is a common term in the security domain, as illustrated in the quotes below:

"Not all crowded places will share the same risk profile or have similar vulnerabilities, so the principle of proportionality should generally be applied to any prevention-related activities. This means that protective security measures not

only need to be proportionate to the level of assessed risk, they should also strike a reasonable balance between protecting the public and, where possible, preserving the public's use and enjoyment of these places. When measuring proportionality it should be recognised that prevention and mitigation activities related to terrorism may also provide broader crime prevention and public safety benefits. (Australia -New Zealand Counter-Terrorism Committee, 2017)

"Protective security measures should be proportionate to the level and type of threat." (NACTSO, 2017)

"All corrective and preventive actions taken to eliminate the root cause of the hazard must be proportionate to the magnitude of the incidents it may cause." (The Supreme Council for National Security, 2016)

Proportionality is not a crystal clear term and often begs the question of how to measure it and how to compare a forecasted uncertain problem with a current action. In some cases, such as with insurance, the proportionality can be measured where the premium paid is proportional to the probability of the risk eventuating and the damage done. When used in security or other risk domains, the term is imprecise, relies on common sense or "what a rational person would do". If the people involved have different ideas of common sense or what is rational, it definitely needs solid qualification.

Local optimisation

A concert with ten thousand people dancing, talking, calling out, singing along, moving around the site can seem confusing. The noise alone is disorientating. Risk response decision must be made in the confusion of a crowd at an event. Not all the information needed to make the best decision is to be found in this of state of affairs. Example 2.1, Moving bollards, illustrates this type of situation.

However, a decision must be made and action taken. To assess the situation quickly using experience and knowledge is known by the technical term 'local optimisation'. The person is assessing the choices with the limited surrounding knowledge and their own experience, and optimising, or making the best choice possible. The person must also know when the information they have analysed is sufficient to take action. This is quite different to having time to plan and consider options and consult with stakeholders. It cannot be ignored and is essential in working on events and in crowded places. Similar conditions of stress, noise, and general confusion is found in frontline war, policing, disasters and emergencies. All of these need, at some time, a person to make the best decision at the time. Taking no action at all is a far greater risk.

Example 2.1: Moving bollards

The new security environment for public events means there are now bollards or cement blocks on the streets. They are placed there as a temporary measure to prevent hostile vehicle attacks or at least minimise their damage. These fixed barriers contrast with the need for crowd management to be responsive to conditions as they unfold. Moving barriers is an important part of controlling the flow of the crowd. Moving bollards, on the other had requires at least a forklift and the personnel to keep people away from the machine as it works. Also there are often a large number of bollards. Hence there will be quite a few forklifts. As it is dangerous to mix vehicles and pedestrians, the forklifts must be placed away from the crowd but near enough to safely drive to the spot. Who will make the last minute decisions to move the bollards? Who has the experience in hostile vehicle attacks to make an informed decision and override the security plan?

The OODA loop, described in Chapter 5, is a useful tool in dynamic, time critical and risky circumstances. It was developed to enable quick decisions for fighter pilots. The acronym means

♦ **Observe**: this requires situational awareness

♦ **Orient**: the part of gathering information, experience,

♦ **Decide**: which may be local optimisation

♦ **Act**: with an appropriate response.

Triggers

Triggers are more than a warning sign. They initiate an action to minimise a potential major problem. Triggers are particularly useful with events and crowds as problems can escalate very quickly if not 'nipped in the bud'. The question of whether a small issue or a warning will accelerate to a disaster is one answered by experience and expert knowledge. For example many people will know the warning signs of a storm approaching and that will trigger actions to seek shelter. A sudden wind change may be the warning for a large storm and this needs more than just keeping out of the rain. A detailed plan may be put into action including covering all the electrical equipment, fastening all the marquees, checking any electrical cords and their connections. It could include cancelling the event and evacuating the site. Example 2.2 illustrates that an evacuation plan is not a simple answer.

Example 2.2: Evacuate into trouble

Evacuating the site may not solve the problem. At one concert in a vineyard a storm suddenly appeared. Leaving aside the fact that 'suddenly appeared' is a value laden term and that storms are common at that time of the year and can grow very quickly, the event team evacuated the site. This meant getting everyone off the site to the exposed dark streets and lanes surrounding the vineyard in a remote rural area. The vineyard itself had ample shelter from wind, rain and lighting, but that was not used in the plan to 'evacuate'.

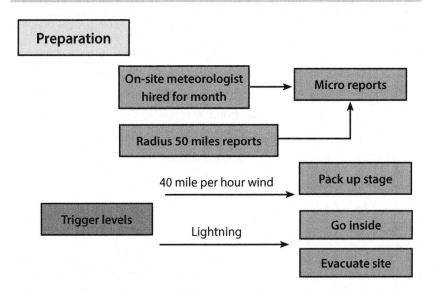

Figure 2.1: Trigger process for a festival

Figure 2.1 shows the process for one event in the USA Mid West where lightning storms are prevalent. The risks is so high that the preparation involves an on-site meteorologist. This preparation is combined with deciding on levels that will trigger actions. The 40 mile an hour wind is a trigger to pack up the stages and evacuate or, in the case of lightning as well, invacuate the site.

The *Safe and Healthy Crowded Places* manual (AIDR, 2018) gives an accurate explanation of trigger points:

Trigger points based on crowd considerations may be used to identify when escalation of management plans is required at a crowded place. Triggers may occur in response to common pre-defined occurrences – e.g. the end of an event or a band's set list – or in response to emerging hazards and threats, e.g. flow stoppages. Triggers may have different thresholds depending on the type of event and crowd present.

Trigger points for escalation of responses related to crowd considerations may include a change of these factors:

◆ *density,*

◆ *change of mood, including significant increase in humidity,*

◆ *flow stoppage,*

◆ *alcohol and intoxication,*

◆ *end of the event, or half-time,*

◆ *setlists, e.g. when stages close.*

As well as density and flow rates, crowd mood is an indicator of possible problems. All these issues should be considered when there is great stress on the crowd.

According to the comprehensive *Guidelines for Concerts, Events and Organised Gatherings* (WA, 2009) "*It is imperative that prior to an event a protocol with clearly identified triggers is developed to enable events to be stopped in a timely manner. This is an extremely important issue that is rarely in place.*"

A warning sign is an indication of an impending hazard, and a trigger indicates the beginning of a problem. Being able to recognise the trigger and take action may address the problem before it increases in severity. Triggers are identified during the risk management process, and the plans to minimise the risk are then drawn up.

Resilience

Resilience is generally defined as the ability to bounce back after adversity within a short time. The key measure of resilience after the disaster is the time period it takes to return to normalcy. It was mostly used in engineering and the metaphor applied to the psychology, ecology and the community sphere. For businesses it is how long the business takes to return to normal, also called *'business continuity management'*. For crowds and events it is the ability to return to normal after a disruption.

Resilience is regarded as a positive quality, and hence the next question is how is it recognised and enhanced. Researching and measuring factors that contribute to a crowd's resilience have the same problems as measuring risk in general. Controlled experiments will miss many of the factors such as the reality of the disaster. Historical records suffer from a lack of observable factors, uniqueness of the situation and the various confirmation biases such as silent evidence. Add to this the complexity of crowds, free agency and emergent characteristics. Hence resilience planning may be useful but it is not a precise science.

Nassim Taleb's concept of antifragility is the ability to strengthen as a result of stress or crisis. It includes learning from mistakes. Hence a risk that eventuates can be used as a lesson. The problem with risk management is that if it applied, the result is difficult to measure.

Conclusion

"Kein Plan überlebt die erste Feindberührung" "No plan survives contact with the enemy" Field Marshal Moltke, *On Strategy*, 1871

For many years planning has been seen as the cure-all for management problems. When an unexpected situation occurs, the advice has been 'The planning must be improved'. This is a truism, as the only proof that planning didn't work is the situation that occurred. Even with all the planning tools now available there are still huge disasters. Planning assumes a linear progression, stability, plenty of time and the ability to forecast. With crowds this is not always the case. When the plan doesn't work, the event team must rely on the competency of themselves, their staff and the people at the event in the crowd. In the chapter we looked at the actual situation, on the ground and the tools and techniques that help in the actual situation. These are drawn from many areas including military and disaster management. The reader will find many indirect and direct references and examples of these in the proceeding chapters.

Questions , ideas and exercises

Discuss the difference between a system that is resilient and one that is antifragile.

How can the staff be trained to be focussed and at the same time situationally aware? The problem of tunnel vision or fixating with airline pilots was solved by training pilots to scan the instrument panel. Is this a solution in crowded situations?

Section One
Risk Management

Dr Jason Brown describes the basic process to all crowd management planning; assessing risk to enable decisions to direct actions. It is risk management.

As the National Security Director for Thales and Chairman of the Board at Security Professionals Australasia, Dr Brown is also the Chair of the Technical Committee TC262 Risk Management, ISO - International Organisation for Standardization.

ISO/TC 262 Risk management was created in 2011 as a Project Committee and was converted to a full Technical Committee in 2012. Its scope is "Standardization in the field of risk management." The capstone standard is ISO 31000:2009.

ISO 31000:2009, Risk management – Principles and guidelines, sets out principles, a framework and a process for managing risk. It can be used by any organisation regard-less of its size, activity or sector. Using ISO 31000 can help organisations increase the likelihood of achieving objectives, improve the identification of opportunities and threats and effectively allocate and use resources for risk treatment.

ISO 31000 is the national standard in the majority of the G8 and G20 as well as BRIICS economies and is highly valued as a guidance document by a wide range of ISO TCs and their National Mirror Committees as well as a number of United Nations agencies.

Well versed in theory and practice, Dr Brown gives the basic terminology necessary to recognising, thinking, deciding and acting in the crowd management sphere.

3 Risk and Management

In the opening chapter it was argued that the perspective this book takes on crowds is from the complexity theory. A crowd is complex as each individual is a free agent and, in a crowd, responds to the people near them. Instantaneously, their attitudes and behaviours affect that individual. Hence it is the relationships between the people that create crowd behaviour. This is beyond complicated, because the causes and effects are immersed in multiple behaviours and attitudes that are moving and developing. Managing risks in this environment requires models and processes that push traditional management processes to the boundary. Before moving onto the methodologies for managing risk it is worth noting some of the drivers.

The 21st Century multi-polar, hyper-connected, social media driven, fast paced world in which we live presents an environment in which we need to understand and master complexities and uncertainties on a scale never before encountered by the average person.

Business models from the most sophisticated in banking to the most rudimentary in transportation in the developed world and financial services in the underdeveloped, are being disrupted as never before. Survival today requires:

♦ Agility and flexibility, and

♦ Automation.

The Global Financial Crisis of 2008, the international failure to achieve strategic objectives in Afghanistan, Iraq and Syria, all powerfully demonstrated complex interdependencies which were not understood by even the most sophisticated organisations and countries.

This changing risk landscape, our understanding of the nature of risk, the art and science of choice, lies at the core of our modern economy. Every choice we make in the pursuit of objectives has its risks. As we seek to optimize a range of possible outcomes, decisions are rarely binary, with a right and wrong answer. Organisations encounter challenges that impact reliability, relevancy, and trust. Stakeholders are more engaged today, seeking greater transparency and accountability for managing the impact of risk while also critically evaluating leadership's ability to crystalize opportunities. Even success can bring with it additional downside risk—the risk of not being able to fulfil unexpectedly high demand, or maintain expected business momentum, for example. Organisations need to be more adaptive to change. They need to think strategically about how to manage the increasing volatility, complexity, and ambiguity of the world, particularly at the senior levels in the organisation and in the boardroom where the stakes are highest.

Defining some terms associated with risk

In an episode of Star Trek Voyager dealing with trans temporal issues, the First Officer of the USS Relativity (a ship from Voyager's possible future guarding the timelines) says to the captain, demanding a risk free solution, "uncertainty is part of the equation". It makes the point that no decision is able to be made without consideration of the elements affecting the judgement and the potential outcome for your objectives and therefore potential consequences cannot be completely known. The more complex the system in which you operate, the more likely that you will have unknowns that will impact on your decisions and therefore on your objectives.

Donald Rumsfeld is often quoted for his statement, *"There are known knowns; there are things we know that we know. There are known unknowns; that is to say, there are things that we now know, we don't know. But there are also unknown unknowns; there are things we do not know we don't know"*.

This issue of "knowing" was considered well before his statement.

In epistemology and decision theory, the term 'unknown unknown' refers to circumstances or outcomes that were not conceived of by an observer at a given point in time. The meaning of the term becomes more clear when it is contrasted with the 'known unknown', which refers to circumstances or outcomes that are known to be possible, but it is unknown whether or not they will be realized. The term is used in project planning and decision analysis to explain that any model of the future can only be informed by information that is currently available to the observer and, as such, faces substantial limitations and unknown risk.

This is as true in Quantum Physics as it is in a decision to cross the roads or buy a car. At best you can infer the outcome of your decision

Heisenberg, in his uncertainty principle paper, 1927, said in the sharp formulation of the law of causality: *"if we know the present exactly, we can calculate the future"*. It is not the conclusion that is wrong but the premise.

One should note that Heisenberg's uncertainty principle does not say *"everything is uncertain."* Rather, it tells us very exactly where the limits of uncertainty lie when we make measurements of sub-atomic events.

These so-called indeterminacy relations explicitly bear out the limitation of causal analysis, but it is important to recognize that no unambiguous interpretation of such a relation can be given in words suited to describe a situation in which physical attributes are objectified in a classical way. (Bohr, 1948)

Terminology used in risk management

In many cases – especially in emergencies when time is critical – efficient communication is crucial. Correct usage of terminology underpins the effective communication of risk to stakeholders and is vital to risk management. Language should be exacting, and terminology universal.

Terms must be common to managers and staff and should clearly and unambiguously describe problems to stakeholders, such as the emergency services agencies.

ISO 31000 provides a common language for risk management, and this nomenclature has been adopted by governments, emergency services and health services. See Chapter 4 for more on communication.

Uncertainty

A situation in which something is not known or not certain. This includes factors and influences, both internal and external, that make it uncertain as to whether, when and the extent to which they will achieve their objectives.

Vulnerability

The characteristics and circumstances of a community, system or asset that make it susceptible to the damaging effects of a hazard.

The extent to which a community, structure, service or geographic area is likely to be damaged or disrupted by the impact of a unique hazard, regarding its nature, construction and proximity to hazardous terrain or a disaster-prone area. (NERAG, 2015)

Event

The occurrence or change of a unique set of circumstances.

An event can have one or more occurrences, causes and consequences. An event can also be an expected occurrence that does not happen, or something unexpected that does. Note that the term 'event' can be confusing in the field of event management. 'Incident' may be a better term.

Hazard and threat

In the paper by the International Risk Governance council they distinguish risks from hazards. Hazards describe the potential for harm or other consequences of interest. These potentials may never even materialise if, for example, people are not exposed to the hazards or if the targets are made resilient against the hazardous effect (such as immunisation). In conceptual terms, hazards characterise the inherent properties of the risk agent and related processes, whereas risks describe the potential effects that these hazards are likely to cause on specific targets such as buildings, ecosystems or human organisms and their related probabilities.

According to the United Nations Office for Disaster Risk Reduction it is a process, phenomenon or human activity that may cause loss of life, injury or other health impacts, property damage, social and economic disruption or environmental degradation.

Hazards may be multi-hazard, biological, environmental, geological, meteorological, technological. (UNISDR, 2017)

Controls

A measure that maintains or modifies risk, e.g. controls for security risk are alarms, fencing, lighting and accreditation zones.

Controls include any process, policy, device, practice or other conditions and actions that modify risk.

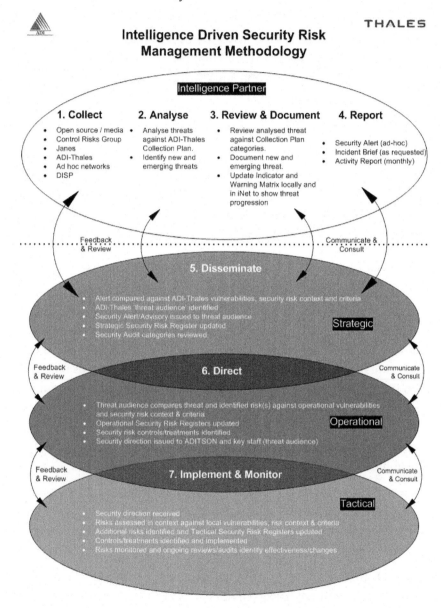

Figure 3.1: Intelligent Driven Security Risk Management Methodology, courtesy of Thales

Mitigation

The lessening or minimising of the adverse impacts of a hazardous event.

The adverse impacts of hazards, especially natural hazards, often cannot be fully prevented, but their severity can be substantially lessened by various strategies. Mitigation measures include engineering techniques and hazard-resistant construction, as well as improved environmental and social policies and public awareness. (UNISDR, 2017)

Likelihood

The chance of something happening, whether defined, measured or determined objectively or subjectively, qualitatively or quantitatively and described using general terms or mathematically (such as probability or a frequency over a given time).

Consequences

The outcome of an event (incident) that affects objectives, e.g. the result if the incident happens.

A consequence can be certain or uncertain, positive or negative, direct or indirect. 'Insignificant', 'minor', 'moderate', 'major,' and 'catastrophic' are the commonly used terms.

Stakeholder/interested parties

A professional or organisation that can affect, be affected by or perceive themselves to be affected by a decision or activity.

The risk equation needs to be considered at each of the two stages of risk management and can be summarised by the following equation, Risk is a consequence ($\vDash$) of the conjunction ($\land$) of Vulnerability and Threat/Hazard.

Architecture model

Business architecture is widely defined as the 'blueprint that provides a common understanding of the organisation and is used to align strategic objectives and operational demands' and build organisational resilience. The business architecture should be viewed as a high-level, best practice approach towards developing a standard for Protective Security Management.

A good analogy for understanding business architecture comes from the building architecture industry, where the architect must have a fundamental understanding of the purpose of the building. More recently the comparison has been to a Town Planner, where the business architecture acts as a liaison between the developers, regulators and the client.

Business architecture is developed by defining the objectives, strategies, processes and requirements of the organisation. It should outline the aims and areas of responsibility for the organisation and also include clear measures and metrics. It is important to limit the scope of the business architecture to the organisational team but to also include the objectives of the stakeholders. Having a clearly defined business architecture will set the direction and priorities for the way the framework is structured and the functions of the organisation.

Designing a business architecture for Protective Security Management requires a clear understanding of the current patterns and the evolving risks and threats in the security environment. It also requires an understanding of how the evolving threat environment determines the operational requirements, the tactical responses and ultimately whether the aims are achieved.

Risk management in organisational architecture

Good risk management can result in value creation and contribute to sustainable organisational performance that helps organisations achieve their strategic objectives. It must permeate all levels of an organisation and hence we need to address Risks at an enterprise level. In the case of security there are a range on malicious actors who will endeavor to achieve their own goals at the expense of the enterprise and often society as a whole. Security professionals must ensure that the risks they seek to control and mitigate are adequately addressed throughout the enterprise hence an understanding of Enterprise Risk Management is important.

We live in a world that is volatile, uncertain, chaotic and ambiguous. These elements collectively feed the uncertainty which plagues decision makers seeking to get the best results. Organisations of all sizes and types need to recognize and manage the uncertainty present in the world and impacting all organisational activities, in order to optimize performance. Developing insight into how risk may affect the objectives of organisations and their stakeholders, and managing this well, is a key performance factor that contributes to the success of the organisation. Good risk management is an expectation of all stakeholders, internal and external.

To build sustainable performance, organisations have to clearly understand the relationships and links between their objectives, strategies and risk management processes in the interests of all stakeholders.

The board, or its equivalent such as the event organising team, has overall accountability for the management of risk. The management of risk should be integrated across all the activities of an organisation. It is a responsibility of all employees. Legislation, regulation and governance codes expect appropriate evidence of responsible risk management processes being applied and managed across all areas of the organisations activities to be available on an on-going basis with clear communication on policies, procedures and processes expected.

Jack Rush, QC, counsel assisting the Royal Commission into the Victorian bushfires said:

> "We submit that leadership cannot be divorced from command. Command does not necessarily involve the issuing of orders or directions, or swooping in to take over at an incident control centre. Command demands a presence: to inform, and if necessary, reassure and inspire. But also to oversight and monitor to ensure that key objectives are being met by subordinates. Leadership and command is not exercised by retreat to so-called co-ordination or to broad oversight. Leadership and command is not exercised by being available if necessary at the end of a telephone." (Victorian Bushfires Royal Commission final report 2009)

Leaders at all levels make decisions with the aim of achieving their objectives or those of the organisation to which they belong. Inherent in any decision is the taking of actions to achieve outcomes where the degree of certainty about the relevant variables that may affect the outcome may be unknown or partially known.

Many people use the term 'risk' to refer to potentially negative outcomes but even when an opportunity to achieve objectives is evident there is still a requirement to manage risk arising from the actions that you choose to take or dismiss.

In this sense risk is neither positive nor negative – risk is what occurs when there is some uncertainty whether you will achieve your expected results, or there are factors that may affect your expected outcome that may not be fully understood or controlled. These uncertain factors may have either a positive or negative effect. For example you choose a site for a crowd event, there are opportunities such as a nice arena shaped slope, a flat area for a stage, that captures the afternoon sun on a winter's day etc.; there are hazards such as narrow road entry with gravel, isolation from response agencies, upstream water course issues that could be affected by heavy rain. You weigh up all these and other issues balancing the opportunity against the potential hazards or threats, put in controls and then take the risk to achieve your desired outcome.

The management standard ISO 31000 provides a risk management that supports all organisational activities including decision making at all levels of the organisation. The ISO 31000 framework and its processes should be integrated with management systems to ensure consistency and the effectiveness of management control across all areas of the organisation.

All sizes and types of organisations need be able to identify, assess and manage the uncertainty that may affect their activities in order to optimize performance. Developing integrated management processes that provide insight into how risk may affect the achievement of the objectives directly contributes to the performance and overall success of the organisation.

Integrating risk management concepts into security practice areas not only provides strong rationale for the security measures but also enhances the efficiency and effectiveness of the chosen mitigations. At the International Standards Committee on Societal Security, it is fair to say that ISO 31000 underpins the management systems that are being developed across the spectrum of standards to sustain and protect our society.

Security professionals today can talk the language of the corporation when they integrate their disciplines with the needs of companies to reduce risk. The risk assessment and the recommendation of appropriate mitigations underpin the capacity of a security professional to explain the difference they make to those with no security background but other responsibilities, such as finance and property management.

Risk classification systems

An important part of analysing a risk is to determine the nature, source or type of impact of the risk. Evaluation of risks in this way may be enhanced by the use of a risk classification system. Risk classification systems are important because they enable an organisation to identify accumulations of similar risks. A risk classification system will also enable an organisation to identify which strategies, tactics and operations are most vulnerable. Risk classification systems are usually based on the division of risks into those related to financial control, operational efficiency, reputational exposure and commercial activities. However, there is no risk classification system that is universally applicable to all types of organisations.

Figure 3.2 illustrates a classification of risk according to their surrounding origin, called *induced*. The table demonstrates how the origin will determine the strategy and the tools and techniques used to treat the risks.

This may be especially true for organisations operating in the public sector and those involved in the delivery of services to the public. There are many risk classification systems available and the one selected will depend on the size, nature and complexity of the organisation. ISO 31000 does not recommend a specific risk classification system and each organisation will need to develop the system most appropriate to the range of risks that it faces.

Enterprise Risk Management—Integrating with Strategy and Performance provides a framework for boards and management in entities of all sizes. It builds on the current level of risk management that exists in the normal course of business. Further, it demonstrates how integrating enterprise risk management practices throughout an entity helps to accelerate growth and enhance performance.

Knowledge Characterisation	Management Strategy	Appropriate instruments	Stakeholder Participation
1 'Simple' risk problems	Routine-based: (tolerability/ acceptability judgement) (risk reduction)	· Applying 'traditional' decision-making · Risk-benefit analysis · Risk-risk trade-offs · Trial and error · Technical standards · Economic incentives · Education, labelling, information · Voluntary agreements	instrumental discourse
2 Complexity-Induced risk problems	Risk-informed: (risk agent and causal chain)	Characterising the available evidence · Expert consensus seeking tools: o Delphi or consensus conferencing o Meta analysis o Scenario construction, etc. · Results fed into routine operation	Epistemological discourse
	Robustness-focussed: (risk absorbing system)	Improving buffer capacity of risk target through: · Additional safety factors · Redundancy and diversity in designing safety devices · Improving coping capacity · Establishing high reliability organisations	
3 Uncertainty-induced risk problems	Precaution-based: (risk agent)	. Using hazard characteristics such as persistence, ubiquity etc. as proxies for risk estimates Tools include: · Containment · ALARA (as low as reasonably achievable) and ALARP (as low as reasonably possible) · BACT (best available control technology), etc.	Reflective discourse
	Resilience-focussed: (risk absorbing system)	Improving capability to cope with surprises · Diversity of means to accomplish desired benefits · Avoiding high vulnerability · Allowing for flexible responses · Preparedness for adaptation	
4 Ambiguity-induced risk problems	Discourse-based: ·	· Application of conflict resolution methods for reaching consensus or tolerance for risk evaluation results and management option selection · Integration of stakeholder involvement in reaching closure · Emphasis on communication and social discourse	Participative discourse

Figure 3.2: Classification from induction

The risk management process applied to security involves the systematic application of policies, procedures and practices to the activities of communicating and consulting, establishing the context and assessing, treating, monitoring, reviewing, recording and reporting risk.

The dynamic and variable nature of human behaviour and culture should be considered throughout the risk management process.

General Risk Management using ISO31000

To create enduring value requires companies to implement risk management strategies based on valid data and sound science and to consult with interested and affected parties in the identification, assessment and management of social, health, safety, security, environmental and economic impacts. This is to ensure that risks are comprehensively reviewed and stakeholders are kept informed. ISO 31000 goes further and lists principles for effective risk management that should be reflected in organisational risk management frameworks.

'Risk' is not an easy term to define, so there are various definitions. While risk is often considered in terms of the likelihood of something happening and the severity of the outcome, risk as a concept is more complex, and that complexity needs to be understood.

AS/NZS ISO 31000 defines risk as the '*effect of uncertainty on objectives*', where uncertainty may relate to a deficiency of information, understanding or knowledge of an event, its consequences or its likelihood. When making decisions as part of managing risk, it is important to remember that this is not an absolute science; it is about managing uncertainty to achieve objectives that may include social, environmental and economic objectives. Risk is also circumstance-specific and has to be dynamic, iterative and responsive to change.

Such uncertainties may relate to technical and human factors; environmental impacts; social benefits; economic factors and political risks. To manage risk effectively, uncertainty and unpredictability must be recognised and information gaps filled to reduce uncertainty. In addition to comprehensive technical work, this requires engagement with relevant stakeholders who will have different perceptions of uncertainty.

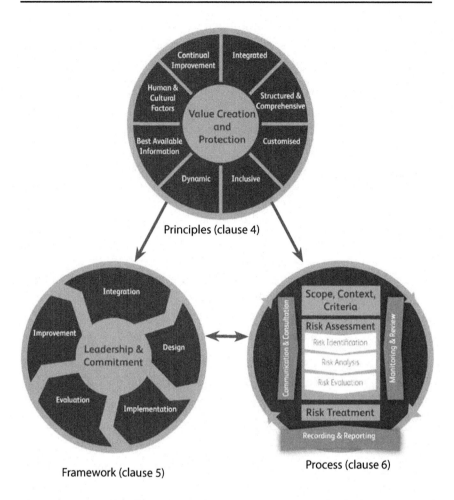

Figure 3.3: Principles, frameworks and process from ISO 31000

Managing risk to achieve objectives in a crowd situation is readily based on the principles, framework and process of ISO 31000 and outlined in this document as illustrated in Figure 3.3. These components might already exist in full or in part within the organisation, however they might need to be adapted or improved so that managing risk is effective, efficient, and consistent.

Applying ISO 31000 to the decision process

General

The risk management process applied to crowds involves the systematic application of policies, procedures and practices to the activities of communicating and consulting, establishing the context and assessing, treating, monitoring, reviewing, recording and reporting risk.

Managing crowded related risk should be an integral part of management and decision making and integrated into the structure, operations and processes of the organisation. It can be applied at strategic, operational, program or project levels.

The dynamic and variable nature of human behaviour and culture should be considered throughout the risk management process. Although the risk management process is often presented as sequential, in practice it is iterative.

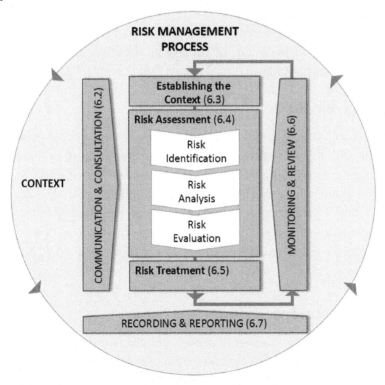

Figure 3.4: Process from ISO31000

Communication and consultation

The purpose of communications and consultation is to assist relevant stakeholders in understanding protective crowd related risk, the basis on which decisions are made, and the reasons why particular actions are required. Close coordination between the two should facilitate factual, timely, relevant, accurate and understandable exchanges of information, taking into account the confidentiality and integrity of information as well as the privacy rights of individuals.

Establishing the context

Defining the purpose and scope

The organisation should define the purpose and scope of its risk management for crowds.

Internal and external context

The internal and external context is the environment in which the organisation seeks to define and achieve its objectives arising from protective security activities.

Defining risk criteria

The organisation should specify the amount and type of protective security risk that it may or may not take relative to objectives. It should also define criteria to evaluate the significance of the risk and to support decision making processes. Risk criteria should reflect the organisation's values, objectives and resources and be consistent with policies and statements about risk management. The criteria should be defined taking into consideration the organisation's legal, regulatory, all other obligations, and stakeholder views.

Risk assessment

Risk assessment is the overall process of risk identification, risk analysis and risk evaluation. Risk assessment should be conducted systematically, iteratively and collaboratively, drawing on the knowledge and views of stakeholders. It should use the best available information supplemented by further enquiry as necessary.

Risk identification

The purpose of risk identification is to find, recognize and describe risks associated with protective security that might help or prevent an organisation achieving their objectives. Relevant, appropriate and up-to-date information is important in identifying risks.

Risk analysis

The purpose of risk analysis is to comprehend the nature of risk arising from crowd behaviour and goals and its characteristics including, where appropriate, the level of risk. Risk analysis involves a detailed consideration of uncertainties, risk sources, consequences, likelihood, events, scenarios, controls and their effectiveness.

Risk evaluation

The purpose of risk evaluation is to support decisions. Risk evaluation involves comparing the results of the risk analysis with the established risk criteria to determine the significance of risk.

Decisions should take account of the wider context and the actual and perceived consequences for internal and external stakeholders. Decisions should be made in accordance with legal, regulatory and other requirements.

The outcome of risk evaluation should be recorded, communicated and then validated at appropriate levels of the organisation.

Risk treatment

The purpose of risk treatment is to select and implement options for addressing risk.

Risk treatment involves an iterative process of:

- formulating and selecting risk treatment options;
- planning and implementing risk treatment;
- assessing the effectiveness of that treatment; deciding whether residual risk are acceptable; and
- if not acceptable, taking further treatment.

Selection of risk treatment options

Selecting the most appropriate risk treatment option(s) involves balancing the potential benefits derived in relation to the achievement of the objectives against costs, effort, or disadvantages of implementation.

Risk treatment can also introduce new risks that need to be managed.

Preparing and implementing risk treatment plans

The purpose of risk treatment plans is to specify how the chosen treatment options will be implemented so that arrangements are understood by those involved and progress against the plan can be monitored. The treatment plan should clearly identify the order in which risk treatments should be implemented.

Monitoring and review

The purpose of monitoring and review is to assure and improve the quality and effectiveness of process design, implementation and outcomes. Ongoing monitoring and periodic review of the risk management process and its outcomes should be a planned part of the risk management process, with responsibilities clearly defined.

Recording and reporting

The risk management process for protective security and its outcomes should be documented and reported through appropriate mechanisms.

Decisions concerning the creation, retention and handling of documented information should take into account, but not be limited to, their use, information sensitivity, and internal and external context.

4 Crowd Risks and Advanced Tools

The objectives of organisations engaged with crowd situations are subject to serious threats that can disrupt and compromise their integrity and prevent desired outcomes. By employing a risk management process, organisations can identify and assess threats, vulnerabilities and weaknesses and also the likelihood and consequences to the security environment.

Risk management involves identifying and assessing threats to the operating environment or strategic goals and prioritizing resources to minimise, monitor and control the impact of events (incidents). There are three key areas for assessing security related risks:

1 Information asset assessment

2 Threat assessment

3 Vulnerability assessment

Information assessment includes identifying physical and intangible assets and assessing the value of maintaining the ongoing cost or damages arising from loss. *Hazard/threat assessment* includes identifying potential threats, targets and the likelihood of occurrence. Potential threats include fraud, theft, loss of infrastructure and malicious acts. In assessing *vulnerability*, organisations need to understand the potential inadequacies and weaknesses. Threats differ across organisations and frequently change.

In an organisation affected by or controlling a crowd situation, risk management relates to a specific culture of processes that are engineered towards maximizing the objectives of the organisation. In managing the risks the organisation should take into account the function and objectives, as well as everyday operations. Organisation should seek to implement a positive risk culture and embed risk management practices into the day-to-day activities.

In developing a risk management process for crowd response and management, organisations should identify the following areas:

◆ Risks to people, information and assets and consequential impact on objectives, capability and preferred outcomes.

◆ The acceptable level of risk.

◆ Appropriate protections to reduce or remove risks.

◆ Assumptions – threats (arising from the intentions and capabilities when acted on), vulnerabilities, consequences, likelihood of occurrence.

◆ Responses and strategies of managing risk.

By adopting a risk-based approach to managing crowd-related risk, organisations can assess and prioritize activities and allocate resources to suit their requirements. Organisations should seek to implement multi-layered business architecture across the platforms to mitigate crowd related risks.

Organisations can be subject to serious threats that can disrupt and compromise their integrity. By employing a risk management process, organisations can identify and assess threats, vulnerabilities and weaknesses and the likelihood and consequences to the security environment.

The three important publications are:

1 International Electrotechnical Commission, International Standard, ISO/ IEC 31010:2009, First Edition, 2009.

2 Standards Australia/Standards New Zealand Standard Committee, AS/NZS ISO 31000:2009,Risk Management-Principles and Guidelines, November 2009.

3 International Organisation for Standardisation, ISO Guide 73:2009, Risk Management-Vocabulary, First Edition, 2009.

Definition(s) for the security of crowds

Protective security

◆ The framework implemented to identify, respond to and reduce the risk of harm from malicious acts.

◆ The measures to reduce the risk posed by malicious actors.

♦ Processes and activities that protect people, assets and information from malicious acts.

Physical security

Physical security is an essential part of protective security. It is approached from a risk-assessment basis and covers a wide array of assets and elements that require an increased level of protection from physical circumstances that may cause harm. Physical security extends to the protection of personnel, the layout and design of locations and institutions, as well as access to equipment, data, systems and networks.

There are four main components for managing physical security that are widely accepted across both civilian and military organisations, known as DDDR:

♦ Deterrence

♦ Detection

♦ Delay

♦ Response

Recovery can also be included as a last step in the process.

Deterrence is designed to convince the potential attacker that a successful attempt is likely to be thwarted. Deterrence methods usually include physical barriers, walls, fences, signage and lighting. Location and design also play a role in protecting the asset. Alarm systems and guards are also a key element in visibly highlighting the level of deterrence.

Access controls are an important element for deterrence. This includes Personnel ID management, gates, doors, locks and electronic access controls for data protection. Policies and procedures are a requirement for setting the organisations expectations and guidelines.

Detection is an important tool for monitoring physical sites and systems and recording any activity that occurs. Surveillance is essential at every level of the four components.

Delay is preferable, particularly for physical sites, where preventing the perpetrators from completing their act is desirable. Slowing an attack-in-progress can also allow the organisation to respond before assets are compromised.

Response is the actions that are taken by the organisation once they are aware of a breach or potential breach. The response must be well equipped and prepared to deal with the nature of the incident.

Restore is the final step in the process, where the organisation should review the incident, assess the level of security and analyse whether the malicious acts were successful. The outcome will determine whether a change to the level of security is required.

While the above elements are common to physical security, different countries have different applications for the term depending on their priorities. The Australian Government application focuses on the protection of employees, work health and safety, and the physical protection of equipment and assets. The British have also extended the term to include preventive measures for managing other threats such as chemical and biological weapons. For the U.S, the focus appears to be primarily around the protection of information.

Personnel security

People are an organisation's most critical assets, but are also its greatest vulnerability. Personnel security is an important part of the protective security landscape aiming to reduce the risk to people, information and assets from malicious acts. Effective personnel security requires a comprehensive approach that also complements the other areas of protective security management.

Personnel security primarily depends upon security organisations promoting and maintaining a strong security culture, which can foster a level of assurance around the credibility of the organisation. As one industry leader commented: *"While we may have in place physical security measures and state of the art information security, it is the integrity of our staff that is the key to effective security..."*

Security organisations maintain policies and procedures designed to protect their people and safeguard information they hold. Personnel security is comprised of a number of key components including security vetting and clearance procedures, people management and deterrence. All components should be approached from a risk management basis.

Organisations will often focus personnel security on the recruitment and clearance process for new employees, but it is of equal importance, if not greater, to also employ strong personnel security policies for long-standing employees. The 'trusted insider' can be one of the most significant threats to an organisation. Maintaining ongoing security training and education can reduce the risk of insider activity that may be threatening to the assets of the organisation. Good line management and regular appraisals are also a valuable method for early detection of changes in the situation of the employee.

Exit interview and separation processes are part of maintaining a robust personnel security culture. Organisations have a duty of care to properly prepare their security-cleared personnel for life after they have left the security industry.

Advanced risk assessment

The science of risk management has come a long way from the original standard, the AS/NZS 1999. It has drawn from many expert domains around the world. These are as diverse as: information science, data analysis, medicine and the military. All these areas include uncertainty and decision making. For this reason the ISO.IEC 31010:2009 – Risk management – Risk assessment techniques has been added to ISO 31000. The standard includes advanced techniques of risk assessment. Many of these are applicable to the management of crowds. The reader will find some of these tools in the chapters of this textbook. They have arisen from quite different domains and specialities and are highly developed.

Scenario analysis, for example, is common in the military and now is used for the risk planning of major events. Fault tree analysis is a mechanical engineering technique used in such disasters as aircraft failure. However it now is used in many areas such as finding the factors at music events that contribute to death by drugs. The following section gives an outline of some of the tools and their current or possible application to the crowds.

In summary the list is:

- Brainstorming
- Structured or semi-structured interviews
- Delphi method
- Checklist
- Preliminary hazard analysis (PHA)
- Hazard and operability study (HAZOP)
- Hazard analysis and critical control points (HACCP)
- Toxicity assessment
- Structured What If Technique (SWIFT)
- Scenario analysis
- Business impact analysis
- Root cause analysis
- Failure mode and effects analysis (FMEA)
- Fault tree analysis
- Event tree analysis
- Cause and consequence analysis
- Cause-and-effect analysis
- Layer protection analysis (LOPA)
- Decision tree
- Human reliability analysis (HRA)
- Bow tie analysis
- Reliability centred maintenance
- Sneak circuit analysis
- Markov analysis
- Monte Carlo simulation
- Bayesian statistics and Bayes nets
- FN curve
- Risk index
- Consequence/probability matrix
- Cost/benefit analysis
- Multi-criteria decision analysis (MCDA)

Brainstorming

Brainstorming involves stimulating and encouraging free-flowing conversation amongst a group of knowledgeable people to identify potential failure modes and associated hazards, risks, criteria for decisions and/or options for treatment. (ISO 3010)

Brainstorming is commonly used in risk management meetings to ensure that potential problems are identified. The nature of events means there will always be something new such as programs, suppliers, sites and audience demographics. It is here that the human experience and imagination must be allowed to wander and discuss possible situation. What is called "counterfactual" is important at these meetings. These are imagined situations that are the opposite of what would be expected or planned. "What if?" is an important question in a brainstorming session.

The results or outputs of the meeting can then be analysed using the other tools and techniques discussed below. Fundamental to this is the risk analysis outlined in Chapter 3.

An example of a "what if" is – what if the police wanted a cement bollard moved during the event? Some of the questions would be: Do they have the authority? What is required to move it? Will this disrupt the crowd? What is the flow on effect in a crowd at various times during the event? How would the crowd management staff explain this to the police at the time?

Bayesian analysis

Now that we have data from around the world, crowd management and security companies can use certain tools that previously would never work. Bayesian analysis is a high level mathematical set of tools based on probability analysis. The analysis works in an uncertain environment that is complex. Hence the fit with large crowd management is perfect. Data on crowds can be used to find correlations. Within this textbook, such as in the Health section, Chapters 9 and 10, the medical resources required for an event can be planned to an approximate level, if the initial state of certain variables or factors are known. These can be the age of audience, type of program, number of people, alcohol, type of weather and type of terrain. The way these factors will impinge on the medical needs of the audience can be inferred by using Bayesian analysis. It can develop a model. The state of these factors are the inputs to the model and the level of the likelihood of certain injuries will be the output. It is probability based, meaning it is never certain. However it does help with the planning of any event. This area is in its infancy and there is more on this in Chapter 5, Crowd Behaviour Theory, with regard to mood, density and flow of a crowd.

Multi-criteria decision analysis (MCDA)

Surprisingly MCDA is found in many areas of event management. A form of this is often used to decide on sponsorship proposals. A simple version, for example, is used by private event companies to decide on the best hotel to use for an event. The Olympics had a very sophisticated software version to decide on which cities should host the Summer and Winter games. The process is to break any major decision down into its components and options. For example choosing between five venues to host an event. Depending on the type of event we could have these as the criteria: entrance, exits, security, transport, accommodation and so on. Each criterion is given a score for each venue. This is set out in tablature form. By adding up the score for each venue, it will be obvious which one to choose. Of course this is a simple example and there are far more aspects that can be employed. But that is the basis of MCDA. The reasoning behind any decision can be clearly demonstrated by showing the MCDA table if there is a problem in the future. The decision making is traceable and accountable.

Cause effect analysis

A number of the advanced tools come under the heading *causal analysis*. They are represented by diagrams of nodes (or squares) and arrows. The nodes are actions or states and the arrows are the causal relations between them. There are versions of this called Isikawa Diagrams and Effect to Cause. Figure 1.1 *Cause effect diagram for drug deaths* is an example of this analysis. The diagrams assist the event team in working out the often convoluted relationship between an effect and its various causes.

For example a disturbance, such as a fight, in a crowd seated at an event could be the result of people standing and blocking the views of others. But it will be exacerbated by the presence of alcohol and the age group attending the event. The causal model can use counterfactuals, alternatives to what has happened, to assess the strength of the arrows, the relationship between the cause or contributing factors and the effect. If there was a limit to the sale of alcohol, would the fight have happened? The strength of the relationship between the various contributing factors and the effect can be modelled as a probability. Putting all these factors together creates a map of the various probabilities, called a joint probability distribution.

This leads to the use of Bayesian analysis to assess, refine the causal model and therefore predict the behaviour at future events. From the causal model the event management team can allocate the right resources. If it shows that there is a high likelihood of a fight, for example, the event team can hire and train the right security personnel. Causal modelling is a formal term for what most experienced crowd managers would know intuitively after many events. Now we can gather the experience and data from many events large and small around the world. This allows us to test and refine the model.

Other tools and techniques

The list above is large and involved. The question then become which of the methods to use. They are all subject to a cost benefit analysis. The benefits of using any of these tools and techniques must be weighed against the costs. With regard to events and crowds, time is an important factor as there are always deadlines. Even the cost benefit analysis will take time to perform. The reader will find many of these techniques referred to in the following chapters.

For major public events with large crowds, scenario analysis is commonly used. The military call it 'table top exercise'. The advantage of this tool is that it brings many of the stakeholders, such as representatives of the police, emergency services, venues and transport authorities into the one room to work through a set scenario. The scenario could be "a person has a heart attack in the middle of a crowd and the trains stop running". The agencies are each asked what they would do in such a situation and how would they work with the other agencies.

Conclusion

Risk management has undergone a number of significant changes as it has developed and broadened its scope. Financial risk and safety risk were two areas that found common ground in the insurance industry. Engineering risk developed quickly and major projects, such as NASAs Man on the Moon project, created a formal system that became project management. With the arrival of an international standard, risk management is now applied to all aspects of events and crowds. It enables events around the world to be compared and the lessons learned in one area of the world to be applied anywhere. The output of the risks process is a series of plans: risk register, crowd management plan, medical plan and many more. But that is not the end of it. Risk management is a process, a way of thinking and acting. As will be demonstrated in the next section, planning is necessary but not sufficient for crowd management.

Section Two
Crowd Management

If you don't understand the structure of a difficult problem, then you can't solve the problem. (thwink.org)

Andrew Tatrai's section on crowd management combines first-hand experience of many years in front of large crowds, planning crowd management and operations with the theory arising from risk analysis, systems thinking and the science of complexity. This perspective is from the practical use of knowledge with immediate feedback and risk.

Andrew points out that crowds are not a simple mechanical unit that can be engineered and all their behaviour predicted. Each person is an independent agent. The combination of people is a complex of interactions and feedback. As the scale of the crowd increases, new behaviours will emerge. It can happen suddenly and the event management must quickly recognise it and deal with it.

Chapter 5 examines ground breaking work of John Fruin and Keith Stills. He furthers this by introducing a systems thinking approach and Snowden's Cynefin model of complexity and chaos. He concludes by describing his solution to the issues in emergent behaviour through the use of dynamic measurement. By discovering the correct variables to measure, the crowd and event management team can quickly solve the issues of crowds.

Crowds big or small assemble for various reasons. Gatherings can be a spontaneous outburst of human passion for someone in need or plain curiosity. It may be to travel, to watch sport or entertainment, or protest a public concern. Crowds could be attracted by sophisticated communications plans, social media and community, or by being near to others. Crowds have been outlawed, discouraged and admired throughout history. This section considers the knowledge of crowd theory and management. In addition it introduces some new thinking on how to plan for crowd management implementation. It treats all crowds as an asset and considers crowd management as crowd care. Even if a crowd becomes violent, crowd care is still a human instinct for all parties and stakeholders and the crowd process needs to be considered and planned.

5 Crowd Behaviour Theory

Crowds and crowd behaviour are consistently studied in an attempt to make sense of the phenomena that affect human safety. However, crowd deaths and incidents continue to occur frequently, suggesting modern theories around crowd behaviour are not being appropriately understood and applied to crowd management and crowd control. You don't have to have the academic acumen of Alexander E. Berlonghi, the pioneer in event risk management, to agree that without an understanding of crowd behaviour, crowd management and control activities are random, and ineffectual.

This chapter will provide an overview of crowd behaviour theories as a starting point for understanding how they can be utilised to assist in effective crowd control and crowd management.

It is helpful to imagine that crowd theories and crowd models are as diverse as crowds themselves and that as crowds change, evolve and develop, the theories and models must either change with them or the definitions move to a different phase. This text proposes future pathways for crowd management.

There are many collective crowd theories all of them are partially correct in the correct circumstances, none are absolutely complete in providing certainty of theoretical judgement.

What is a crowd?

Stephen Reicher, in his study on psychology (2001), describes crowds as:

"the elephant man of the social sciences. They are viewed as something strange, something pathological, something monstrous. At the same time they are viewed with awe and with fascination. However, above all, they are considered to be something apart"

Crowds have been defined in many ways, they are generally described as a group of people that are close, geographically or logically, and are affected by each other's presence and behaviour. In order to provide a more precise definition, the UK Government sets out five criteria that may jointly identify a crowd.

♦ **Size** – there should be a sizeable gathering of people

♦ **Density** – crowd members should be collocated in a particular area, with a sufficient density distribution

♦ **Time** – individuals should typically come together in a specific location for a specific purpose over a measurable amount of time

♦ **Collectivist** – crowd members should share a social identity, common goal or interest, and act in a coherent manner

♦ **Novelty** – individuals should be able to act in a socially coherent manner, despite coming together in an ambiguous or unfamiliar situation.

Crowds are often labelled as a description of their main characteristic. This is an over-simplification and a dangerous way to create crowd nomenclature, given the same crowd moves between these different types at different stages in its lifecycle.

Common descriptions have been:

♦ **Ambulatory crowd** – A crowd entering or exiting a venue, walking to or from car parks, or around the venue to use the facilities.

♦ **Disability or limited movement crowd** – A crowd in which people are limited or restricted in their mobility to some extent, for example, limited by their inability to walk, see, hear, or speak fully.

♦ **Cohesive or spectator crowd** – A crowd watching an event that they have some to the location to see, of that they happen to discover once there.

♦ **Expressive or revelrous crowd** – A crowd engaged in some form of emotional release, for example, singing, cheering, chanting, celebrating, or moving together.

♦ **Participatory crowd** – A crowd participating in the actual activities at an event, for instance, professional performers, athletes, or members of the audience invited to perform on stage.

♦ **Aggressive or hostile crowd** – A crowd which becomes abusive, threatening, boisterous, potentially unlawful, and disregards instruction from officials

♦ **Demonstrator crowd** – A crowd, often with a recognised leader, organised for a specific reason or event, to picket, demonstrate, march or chant

♦ **Escaping or trampling crowd** – A crowd attempting to escape from real or perceived danger or life-threatening situations, including people involved in organised evacuations, or chaotic pushing and shoving by a panicking mob

♦ **Dense or suffocating crowd** – A crowd in which people's physical movement rapidly decreases – to the point of impossibility – due to high crowd density, with people being swept along and compressed, resulting in serious injuries and fatalities from suffocation

♦ **Rushing or looting crowd** – A crowd whose main aim to is obtain, acquire, or steal something – for example, rushing to get the best seats, autographs, or even commit theft – which often causes damage to property, serious injury or fatalities

♦ **Violent crowd** – A crowd attacking, terrorising, or rioting with no consideration for the law or the rights of other people

Crowd behaviour models

There are many models for explaining crowd behaviour, coming from a variety of fields.

Classic crowd theories

The origins of crowd theory can be traced back to the 1800s, and early thinkers have had a significant impact on the development of the field. In essence, two philosophical schools of thought have dominated the field: *convergence* and *divergence*.

The convergence school of thought evolved from the early work of Le Bon and is based around the idea of 'group mind'. Le Bon stated that every individual in a large gathering is transformed into a crowd member, and as part of the crowd's collective mind they feel, think and act differently than they would if they were alone. This has had a significant impact on the field with several deindividuation-based theories emerging. These deindividuation-based theories focus on describing the process by which a person supposedly loses their sense of individual identity and therefore engages in behaviour that is out of character, and often extreme.

The divergence school of thought suggests that it is common traits in the individuals that make up the crowd that develop group behaviour. This school suggests that similar innate drives that individuals in a crowd supposedly share give rise to similar behaviour. For example, the Social Identity Theory suggests that a person's sense of self is based on their group membership(s), and that people who belong to the same group form an 'in-group' (us) and discriminate against the 'outgroup' (them) to enhance their self-image. The Emergent Norm Theory suggests that new social norms emerge within a crowd as key members of the crowd (leaders) suggest appropriate actions and following members fall in line, forming the basis of the crowd's new norm.

Refuting the classic theories

It is only in the last 20-30 years that these schools of thought have been met with serious criticism and been overtaken by a more scientific approach to crowd analysis. The main problems identified with these classic models are:

Qualitative descriptions

The theories only offer qualitative descriptions, which are insufficient for the behaviour to be consistently modelled and applied to crowds more generally. This is partially because crowd behaviour has traditionally been studied through ethnography, which captures a narrative that can be difficult to model. Qualitative modelling also suffers from human bias and inconsistency, producing unreliable variations.

Computational models

More modern approaches to crowd analysis have begun to use computer generated crowd simulations, for which broader algorithmic models are required. Cellular automata (CA) computational models are effective for simulating physical systems and can capture the essential features of a system where global behaviour arises from the collective effect of simple components that act locally, such as in a crowd. These models however are limited by the degree of the rules that the agents (model of persons) that the system applies. They are also limited in that the simulation cannot mimic or replicate the speed of change in crowd characteristics when certain input factors or panic sets in. Computer modelling is supportive for design and planning purposes, but as yet ineffective in producing the real life modelling of humans with a large set of variable inputs.

Crowd behaviour as different from other human behaviour

All of the classic models treat crowd behaviour as something unique and suggest that people behave differently when part of a crowd than they do in other circumstances.

More recent observations of crowds in action have undermined this supposition. Adang, in his study of what initiates and escalates collective violence, observed dozens of crowd events and identified that people behave in a way that is meaningful to them, and that the behaviour of individuals within a crowd is not observably uniform. His observations have debunked the myth that individuals in a crowd behave differently than in other circumstances.

Schweingrüber and Wohlstein suggest that the classic crowd theories have produced seven myths about unique crowd-based human behaviour that still show up in modern theories and discussions, all of which they have refuted:

◆ **Myth of irrationality** – the idea that individuals in a crowd lose rational thought

◆ **Myth of emotionality** – the idea that individuals in a crowd become more emotional

◆ **Myth of suggestibility** – the idea that individuals in a crowd are more likely to obey or imitate

◆ **Myth of destructiveness** – that idea that individuals in a crowd are more likely to act violently

◆ **Myth of spontaneity** – the idea that in a crowd violence occurs more suddenly

♦ **Myth of anonymity** – the idea that individuals in a crowd feel more anonymous

♦ **Myth of uniformity/unanimity** – the idea that all individual in a crowd act in the same way.

With the idea of 'group mind' abandoned, more general theories of human behaviour can be applied to crowds. This moves the conversation away from some amorphous 'group mind' to the role of the individual in the crowd, and emphasises that understanding the individuals that make up the crowd is necessary to explain crowd behaviour patterns.

That one model fits all situations even though crowds are diverse as humans themselves

Classic thinking has always tried to develop a model to suit the observations. To label and predict outcomes based on what has been observed. The difficulty becomes when new activities are observed which don't fit previous estimations, whereas a modern approach is that the observations develop the model. Data science now allows the data to develop the algorithm.

Focus on extreme crowd behaviours

Classic theories of crowd behaviour have all focussed on when crowds behave in an extreme fashion, such as rioting, or evacuations that lead to crisis. There is still a tendency towards this today as a significant amount of crowd related literature comes from law enforcement and other agencies that are focussed on managing crowds and retrospectively explaining crowd disasters.

The focus on these extreme situations, or 'disorder', has led to overly simplistic explanations being applied to the full range of crowd activities and obscuring the overall working of the system.

Modern principles of crowd analysis

In her review of the current literature around crowds, Wijermans describes several of the current theories on crowd behaviour, all of which have contributed to moving the discussion forward. Three common key principles can be identified from these studies:

Crowd behaviour is generated by individuals

The idea of 'group mind' has been rejected, and it has been accepted that the behaviour in a crowd is generated by the individuals within it. To understand crowd behaviour, the group level needs to be related to the individual level. Crowd behaviour is context dependent.

Every situation and individual is unique, and changes, so is the behaviour that is generated by an individual in a certain situation. Behaviour occurs in a context, and (perceptions of) external stimuli affect an individual's choices in any given context. Both physical and social context are important. For example, an individual will be drawn to different locations depending on whether it is raining or not, and will move differently if they are alone or in a group. It is important to incorporate the relevant situational factors to understand the behaviour patterns of a crowd at any given moment

Crowd behaviour is dynamic

Behaviour in a crowd continuously changes. A crowd may start behaving a certain way, but this behaviour can change rapidly as its physical and social context changes – the behaviour of the crowd will change if something physical happens, such as the outbreak of a fire, or something social happens, such as the home team suffering an unexpected loss

Crowds operating as systems

Accepting these modern principles, suggests that a crowd functions as a system, in which the individual components cause the behaviour of the system as a whole.

In essence systems are groups of elements or 'agents' and processes that function together to produce something that is emergent from the interactions of the different parts of the system. Boundaries separate a system from other systems and from the rest of the universe. A closed system does not allow free transfer of energy or matter across the boundary. An open system allows inputs or outputs across the boundary. The system collects inputs, which are taken either internally from the system or from external influences, and depending on the structure of the system and the processes happening within it, different outputs are produced.

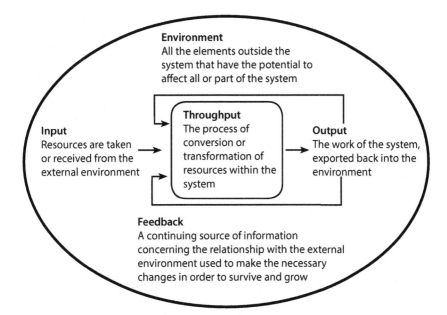

Figure 5.1: An organisation as a system, with inputs, processes and outputs. from Katz and Kahn, 1966.

Systems vary significantly in their complexity. Snowden suggests that systems can be divided into three general types: ordered, chaotic and complex adaptive systems

Ordered systems possess repeating relationships between agents, which display predictable cause and effect relationships. For example, computer systems are ordered systems, in which the results are constrained by the actions of the user and the behaviour of the system is relatively simple to predict from the nature and activity of the elements within the system.

In **chaotic systems** agents are not hindered by any constraints imposed by the system, and the system has a significant number of relevant factors which can be difficult to disentangle. An example of a chaotic system is the field of metrology: the number of elements to be considered when predicting the weather is large (e.g. temperature, humidity, barometric pressure, clouds, sunshine, winds, topography and ocean influences) and small changes in one area can have a significant impact in other areas.

Then there are **complex adaptive systems**. Similar to chaotic systems in their complexity, complex adaptive systems are generally open, which means that they are affected by external factors, and like chaotic systems, a significant number of different elements interact with each other and change according to stimuli. However, what makes these systems unique is that the internal dynamics confound easy description and often defy predictions.

There are seven key elements to adaptive systems, which can also be used to identify crowds as an example of this type of system.

Large number of independent elements

Like chaotic systems, complex adaptive systems are made up of a large number of elements that act independently and interact with one another. As crowds are by definition a gathering of a large number of individuals, crowds clearly meet this criterion.

Wijermans suggests there are three levels of individual agency within a crowd:

◆ **Intra-individual (cognitive)**: an individual's internal world or mental state that interacts with the external world through perception and influences their behaviour.

◆ **Individual (behavioural)**: behaviour arises from cognitive activity, but can be observed, and acts as stimuli in the external world.

◆ **Inter-individual (group)**: group level observable patterns which are emergent or aggregate descriptions of individuals as a whole. This can refer to groups of families and friends, which are a group of individuals, but behave differently from individuals alone.

Unfortunately most studies only look at individuals within a crowd, but individuals move in groups (friends, couples and families) which equal medium scale structures, and their impact on crowd dynamics is not well understood.

Nonlinearity

Specifically, nonlinearity is evident in the interactions between the various elements in the system. Nonlinearity means that there are multiple, densely connected, overlapping feedback loops between the elements that make up the system. This ensures that the system will function in unexpected ways, given the webs of connections among agents, resulting in unpredictability.

The interactions between members of crowds are generally nonlinear and disorganised. Individuals are more likely to exchange information with people close to them, either physically or socially. For example, individuals in a crowd can see how other members of the crowd who are physically close to them are behaving, where they are moving, and how they are moving, and can make decisions about their own movement patterns on this basis. Socially, individuals are more likely to exchange information verbally with people that they already know, either face to face or via their mobile device. As a result, individual members of the crowd do not have the full picture of what is happening across the crowd, but only part of the picture, and make their individual decisions on the basis of this partial information.

Adherence to simple rules

Simple rules influence the interactions between the elements in the system. Agar suggests that some elements in these systems follow simple condition-action rules, where under certain circumstances elements within the system perform a particular action.

James Surowiecki illustrates this in *The Wisdom of Crowds* by describing how a flock of birds accomplishes its group journey by creating a well-ordered formation that emerges from each of the birds following a set of four rules: (1) stay as close to the middle as possible, (2) stay 2-3 body lengths away from your neighbour, (3) do not bump into other birds, (4) if a hawk dives at you, get out of the way. By following these rules the flock is able to self-organise its journey, reach its destination, and handle predators.

Individuals in a crowd generally follow simple, shared rules, largely based on social norms, such as common courtesy, maintaining a certain distance between themselves and strangers and following directional signage, etc.

As well as these kinds of shared social rules, a unique aspect of humans and how they make decisions is that it is based on patterns of previous experience.

Self-organisation

Complex adaptive systems are a large collection of diverse parts inter-connected in a hierarchical manner such that organisation persists or grows over time without centralised control. While not always apparent in these systems due to the irrational, nonlinearity of relationships among the elements, complex adaptive systems self-organise without central authority.

Self-organisation takes place at both the micro and macro level. On the micro-level it occurs when individual elements, with their nonlinear interactions, congeal to produce spontaneous patterns of behaviour owing to the presence of rule governed behaviour. These micro-level constructs can join to form macro-level structures with hierarchal characteristics.

In the absence of leadership, which can be provided by elements in the system such as event organisers or security, crowds will self-organise. This is most obvious in violent crowds in which people who don't necessarily know each other manage to form groups that can riot or loot local businesses, but this is also present in peaceful crowds, such as an audience at a music show that will distribute themselves in a logical way in the viewing area.

Environmentally sensitive

The environment in which self-organisation occurs has a profound impact on the stability of the organisation. Complex adaptive systems are particularly sensitive to the initial environmental conditions of the system, and small changes in the environment can have a significant impact on the system as a whole. This kind of dramatic change occurs when the system self-organises to a point of instability. This phenomenon, the tipping point between stability and change, is referred to as *self-organised criticality*.

The behaviour of crowds can be greatly influenced by environmental changes. An obvious example is the weather conditions, with people behaving differently in the rain, or whether the home football team wins or loses their match. Crowds are also highly influenced by the environment in which they were formed. Have people come together for a street fair or to protest? Do the people in the crowd have common or divergent religious or social beliefs? Have people paid to attend an event or is it a free gathering? Has the event been long planned or is it spontaneous?

Emergent behaviour

Systems thinking definitions have led to considering crowds as Emergent behaviour.

Economist Jeffrey Goldstein provided a current definition of emergence in the journal *Emergence*. Goldstein initially defined emergence as: "the arising of novel and coherent structures, patterns and properties during the process of self-organisation in complex systems".

Emergent behaviour refers to the way complex systems and patterns arise out of a multiplicity of relatively simple interactions. There is a range of definitions about how emergence is applied, however in crowd management the most relevant frameworks to classify crowds are ordered but complicated or complex adaptive. Ordered complicated crowds refers to Snowden's Cynefin framework which requires expert diagnosis because the cause and effect relationships are not always apparent. An example of this are areas that are poorly lit, allowing crowd behaviour to be unaccountable and hence, aggression, criminality or sexual predation may evolve. The cause and effect is that well-lit areas do not support antisocial behaviour but not all practitioners will consider this nor feel empowered to raise or communicate this and pursue fact based management.

In a complex adaptive system the behaviour of the system (its outputs) emerge from the inputs from the system agents, which are in turn influenced by environmental factors, nonlinear inputs from other parts within the system, and the way that the elements within the system self-organise. While the behaviour of the system emerges from these elements is difficult to predict from understanding the individual elements, due to the complex relationships between all the factors. The changing or emergence of crowds can occur not only within a crowd but from one crowd type to another.

As an example of emergence, consider an attendee becoming trapped under a car during an accident at a car rally and several strangers stop to help. No controlling authority is present and confusion reigns as the strangers gather. But as they continue to interact, self-organisation occurs, and a purpose, to lift the vehicle, emerges.

One reason emergent behaviour is hard to predict is that the number of interactions between a system components increases exponentially with the number of components, thus allowing for many new and subtle types of behaviour to emerge. Emergence is often a product of particular patterns of interaction. Negative feedback introduces constraints that serve to fix structures or behaviours. In contrast, positive feedback promotes change, allowing local variations to grow into global patterns.

Learning and adapting

Critical to the success of biologically based complex adaptive systems is the capacity to sense the environment, and thus modify goal-oriented behaviour. This key feature of the system is known as *metis*. Metis enables systems to learn from interactions with the environment and adapt their behaviour accordingly.

Adaption is made possible by learning. Feedback loops carry messages back and forth between elements in the system in a nonlinear fashion. Outputs of some elements are inputs for others, so that what started out as a collection of individual agents turns into a coordinated crowd. This occurs as feedback loops modify system behaviour by delivering constantly changing information about the environment from element to element and back again. Metis, combined with additional inputs from the environment, follow a feedback loop to self-organised elements which adapt their behaviour accordingly.

There are two kinds of feedback, positive and negative, in systems theory. Positive feedback describes the loops tendency to amplify the changes on the system, and negative to dampen.

Add an example of a positive or negative feedback Snowden and Boone have pointed out humans present unique challenges to complex adaptive systems because:

♦ Humans have fluid identities that they switch between and therefore are unstable agents,

♦ Humans make decisions based on past patterns of success/failure, not based on logic/rules,

♦ Humans can in some circumstances change the system in which they are operating.

This identification is important, as it provides those trying to manage and control crowds a basis for understanding how a crowd might behave given its composition, structure and environmental and social inputs.

"At the heart of the concept of systems thinking is that the behaviour of a system is an emergent property of its structure. Not its parts. Each symptom can be traced to a particular aspect of the structure. It follows that if you don't know the structure of a complex social system problem, then you will be unable to solve the problem except by trial and error. For difficult problems this is impossible" (thwink.org).

Applying systems thinking to crowd management

How can systems thinking be applied to crowd management? Crowd management typically refers to the systematic planning and providing guidance for the safe and orderly development of events where large numbers of people come together.

According to Wijermans et al. there are two phases to crowd management:

Preparation phase

Event preparation focusses on planning. Historically this has been known as linear planning or cause and effect. Linear planning is is the planning or scheduling of project management tasks where distance is a significant factor in the project. Linear planning considers not only the time factors of a task but also the location factors. This can never be replaced, underestimated or ignored.

Planning is knowledge management and consultation. The process of documenting the facts and research anticipating the crowd characteristics provides evidence of fulfilling a duty of care. There are many administrative ways to apply planning, however systems thinking helps sort factors as inputs. The way we plan crowd events now is to scribble a mind map with all the inputs we can gather from research and consultation as well as venue and place experience and then like a web start to connect each character to all others and estimate if they are an attractor (creating positive change) or a detractor (creating negative change).

Planning involves anticipating what might happen regarding a crowd in a given context and preparing for it. As such preparation includes designing for the desired crowd behaviour, but also foreseeing potential issues and devising emergency and contingency plans. The resulting plan usually targets the design of the site the crowd will use (entry and exit points, thoroughfares, etc.), supporting technical infrastructure (signage, facilities, etc.), number of assigned personnel, and prescribed operational interventions for dealing with normal as well as anticipated critical situations. The quality of the anticipatory analysis of the likely behaviour of the crowd, as well as the effectiveness of the planned measures, are critical to effective crowd management.

Some automated simulation tools exist to assist in planning for certain types of event. These simulations are generally agent-based and model the anticipated behaviour of each individual and their response to one another and external stimuli. These simulations are particularly effective in modelling the pedestrian movement of crowds in spaces so that the planning and structure of entry and exit routes and main thoroughfares can be stress tested prior to the event. They are currently less effective at modelling social behaviour, and a major problem with current tools is that few of the underlying models on which they rely have been evaluated against real world human behaviour. This is extremely difficult as the available human data is limited, and it is almost impossible to conduct the required controlled experiments because of the difficulty in controlling for variables, as well as the ethical considerations as experiments may place a crowd in danger.

Execution phase

During the event, the situation in a crowd must be continuously monitored, assessed, and appropriate actions selected and taken. Communication is key to this, both among the crowd management team, and between the crowd management team and the crowd itself. A solid command and control structure must be in place.

"To keep management from crossing over into crowd control, one of the most important things to do is correctly access the mood of the crowd" (Abbot & Geddie, 2000).

During the event crowd observation and monitoring enables the assessment of a situation and the detection of potential problems at an early stage, ultimately allowing the selection of appropriate action. Information that is typically monitored includes counts of people in identifiable areas, the space between people, the general crowd mood, signs of distress, pushing or surging, indications of bad temper or excitement, etc.

Further, once a course of action has been selected, the actual action and its consequences must be monitored and evaluated. The process of monitoring, interpreting, predicting and deciding, as well as acting, takes place continuously. They are part of a continuous decision cycle.

This need for constant, dynamic decision making was recognised by military organisations when contrasting challenges of decision making between peace time (bureaucratic operations) and war (dynamic, unpredictable environments). In response John Boyd developed the OODA loop to guide decision making in dynamic and changing environment. Standing for Observe, Orient, Decide and Act, the OODA paradigm offers another method for working in complex and emerging environments.

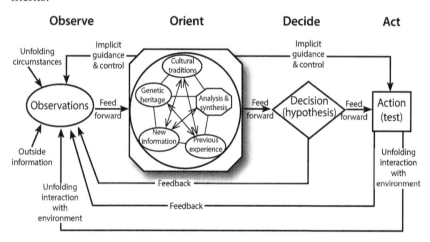

Figure 5.2: John Boyd's OODA loop

Boyd stated that the most important part of the OODA loop is the orientation stage, which relates to the planning stage for crowd management (Snowden, 2008).

Decision-making frameworks

Advice and help for crowd managers in observing and interpreting crowds and deciding how to act can be found in the world of business leadership, as companies and businesses can also be identified as complex adaptive systems, with multiple agents interacting in a variety of ways and subject to the influence of a wide range of external factors.

Management science has until recently been dominated by the concept of order that can be traced back to the early 20th century when Frederick Winslow Taylor developed scientific principles of business management. He developed the idea of seeing business as systems with elements and command and control processes that display clear cause and effect relationships that facilitate decision making. However, over time the complexity of business systems, and the challenge of emergence, have been recognised, and more sophisticated approaches to business management developed.

Snowden and Boone (2007) have been thought leaders in defining how decisions are made in different business environments. They identified that one of the main reasons business leaders fail is the assumption that predictability and order always exists. This assumption that underlies a scientific approach to management is grounded in Newtonian scientific theory, and provides useful simplification, but fails when the challenges faced increase in complexity.

In order to assist business leaders in dealing with the complexity of their organisations, Snowden (2011) developed the Cynefin Framework. 'Cynefin' is a Welsh word that means *"multiple factors in our environment and our experience that influence us in ways we can never understand"*

The Cynefin Framework offers a methodology for sorting challenges into five contexts, defined by the relationship between cause and effect: Simple, Complicated, Complex, Chaotic and Disorder. As a simple overview, Simple and Complicated contexts are when there is a clear relationship between cause and effect that is perceptible, and therefore decisions can be made on the basis of facts. In a Simple context there is only one clear answer (dealing with known knowns), in a Complicated context there are multiple potential correct answers and selecting a response from these adds the complication (known unknowns). Complex and Chaotic contexts lack this clear relationship between cause and effect and decisions need to be based on emerging patterns. In the Chaotic state it is the unknowable, and the relationship between cause and effect is impossible to identify because it is constantly shifting. Disorder is when the context is unclear and different perspectives rise to prominence. Snowden and Boone suggest that in this situation context should be broken down into constituent parts, which can be aligned to one of the other four contexts. It is also noted that organisations can migrate from one system to another by varying inputs.

Crowds are as variable as the humans that make them. Crowds can develop from one form to another. They can evolve from complicated ordered systems for small, calm events to complex systems for mass gatherings with energy and passion. A crowd manager must learn to recognise the type of crowd systems they are working with. In addition a crowd manager must recognise when a crowd can evolve into a different form by way of inputs such as more people or more energy.

A definition of the first four contexts with details of Snowden and Boone's advice on how leaders should respond and potential pitfalls is provided in the table below.

Snowden and Boone	Observations of crowds
It involves a large number of interacting elements	Individual and groups as agents , previous past experiences motivation, density, crowd flow and mood and phycology, size , numbers and groups, Light sound, control forces, weather
The interactions are nonlinear, minor changes can produce disproportionately major consequences	Some inputs can cause panic, reaction to peer influence or police violence. Barriers or blockages cause crush, music and alcohol can exaggerate behaviours
The system is dynamic, referred to as emergence	Crowd behaviour can develop quickly and expectantly with the right catalysts
The system has a history, past integrated with Present	People humans have intentionality, identities &can change the system (rebel against guidance / force) create self-generating leadership for internal groups
May appear ordered but hindsight does not lead to foresight, because conditions continually change	Exactly as crowds seem at times, even mass movements appear ordered till a blockage, a noise, a scare sets the crowd off in a disordered pattern.

Table 5.1: Comparing Snowden's adaptive complex environment to crowd situations. From Snowden and Boone (2007)

The open-systems approach was first applied by Katz and Kahn, who adapted General Systems Theory to organisational behaviour. Although they defined open and closed systems nearly all systems fit into an open systems approach dependent on the sensitivity to the external environment. This fits a crowd situation as the external environment sensitivity is a result of light, sound, entertainment and the behaviour of other parts of the crowd itself.

6 Crowd Planning and Preparation

This chapter reviews planning methods and practices. Significant work has been published and used for long periods on planning methods. Preplanning is essential due to the life safety factors that a crowd can develop *in situ*. Planning can be considered in two phases. Information and background planning essential to communicate facts and identify risk areas in crowd management and operational planning. This then provides resourcing and contingency planning once the operation is in place. Like military operations both phases are important, however in many crowd situations operational and contingency planning is given less scrutiny. This is because the plans are normally scrutinised by authorities, councils, government, venue or land owners and they are more comfortable with pre-information type plans that inform them of the context background and communication flows. How the crowds are managed by security contractors is not usually an area they are experienced in, hence less attention is paid to these areas. The aim of this chapter is to provide enough knowledge for all event stakeholders to review and discuss practical implementation issues in security deployment and control.

Planning and preparation requires an increased focus for crowd management because the emerging behaviour from the collective requires more options to be considered and prepared for. As crowds can cause life safety issues and because agents and systems can interact to exaggerate interactions and responses quickly, preparation and contingency planning is vital. Crowd risk assessments have to be conducted to understand and communicate the magnitude of the problems that can occur. If the consequences of the crowd activity are significant to the risk appetite of the organiser then response methods and measures should be developed and implemented. An example of this would be preparing additional signage, barriers and guards to divert pedestrians away or around potential bottlenecks when the flow becomes too congested.

Theoretical capacity analysis

Many planning documents start with a spatial capacity calculation which estimates available square meters of useable space.

Theoretical capacity is based on a pure physical capacity. Human body shapes and sizes vary with crowd demographics, children, prams, young and old persons. These can never be calculated accurately however a guide to an upper capacity is a good starting point. The UK Green Guide (2018) is the benchmark document which declares the upper limit of persons per square metre is 47 per 10sqm or 4.7 persons per square metre. This is assuming they are all of average size (different nationalities differ in sizes) and they are static. In reality this is not a practical guide, just a notion to prevent any capacities exceeding a theoretical maximum. It is primarily as starting point to ensure emergency egress can be achieved within approved egress times; i.e. assumptions need to be made on egress time that require a total volume number to exit. The Green Guide capacity calculations also only apply to closed or sealed venues and cannot be applied to large free mass gatherings such as fireworks or city light shows where capacity cannot be controlled. Green Guide capacity estimations also do not consider crowd issues when flows exceed gateway or corridor capacity, however this chapter supports the empirical commencement of crowd planning with a capacity statement. It is good practice to start with a quantitative approach to define management process.

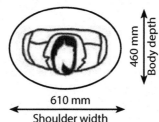

Figure 6.1: Pedestrian operating space and clearances

Body depth and shoulder width are the primary human measurements used by designers of pedestrian spaces and facilities, where shoulder breadth is the factor affecting the practical capacity. The plan view of the average adult male human body occupies an area (the body ellipse) of about 0.14 m2. However, a 460 mm by 610 mm body ellipse equivalent to an area of 0.21 m2 is used to determine practical standing capacity, allowing for the fact that many pedestrians carry personal articles, natural psychological preferences to avoid bodily contact with others and body sway. Figure 6.1 *Pedestrian operating space and clearances* from the Government of Western Australia guidelines illustrates this.

An initial spatial maximum should be four persons per square metre provided this is never used as a final capacity as crowds do not average out spatially. Spatial capacity estimates are of limited value in crowd management situations.

Traditional crowd management planning

Even with the correct average number of people in an area, internal movements through bottlenecks, gateways and pathways can all become dangerous if too many people try to push through them.

DIM-ICE

The DIM-ICE tool was created by Prof G.K. Still in 2001, and referenced from his text book *Introduction to Crowd Science* (2013). It is a breakdown of the functional areas: Design, Information and Management; across the three periods of an event: Ingress, Circulation and Egress. It is derived from the origins of the functional areas involved across the periods of crowd movement in an event. It is essentially a time matrix, of the design factors, information and management opportunities that exist during the three periods of crowd flux. Police, council and host venues appreciate the breakdown as it provides them with the information they need to make resourcing and conditioning decisions. Normally promoted as a spreadsheet comparison, this limitation restricts the detail of information that can be provided, which is a weakness. Additional information has been included by practitioners, including crowds on approach and egress towards transport hubs. This method is still an effective tool to identify crowd risk areas and points over the period of the event.

International risk management standard

Crowd planning from a risk management approach is a consultation approach based on identifying risks and working with groups to implement agreed controls. T he international risk management standard ISO

31000 is applicable to all activities. Crowd management can benefit from this method. A crowd risk management approach requires consultation with all stakeholders to identify the things that can go wrong. Internal, external and internet research should be applied to cover all possibilities and create awareness of possible outcomes. This is the strongest feature of this methodology, because this also builds relationships with stakeholders and begins a working group approach. Risk management requires consultation throughout the process, including at the assessment and control development stage, which is often overlooked. The correct way to apply crowd risk assessment is to ensure that a group of seasoned, experienced and focused stakeholders discuss and agree on all stages of the decision making. Once an educated group is gathered and has considered the answers, risks and controls become easy. This method is now an internationally used process and good practice must include an application of this methodology.

Security risk assessment of crowd flows – what could go wrong?

An understanding of what risks can occur, and why risks precipitate to become incidents, is the fundamental basis of risk management. To determine if event objectives are achievable, detailed risk identification and causation needs to be considered and included. This research needs to include and outline the principles of crowd incident causation from bottlenecks to overcrowding to security interaction and communication with the crowd.

The foundation of all crowd management derives from the work of John J. Fruin (1971). Among his spatial calculations and level of service guides developed for the US highway capacity manual and his work on pedestrian movement he developed a F.I.S.T. model, which details the elements that affect crowds and crowd behaviour. Fruin was the first to clarify crowd risks revolved around:

1   **Force** or energy of the crowd which also included many factors such as crowd size, movement, motivation, peer group leadership and external influences.

2   **Information**, crowd visual perceptions signals messaging and precedent.

3   **Space**, how much physical room was available and the impact it had on other factors.

4 **Time**, the period over which factors materialise will affect judgments and reactions.

In hindsight, the complexity of his model and the large number of possible interactions supports the development of an emergent behavioural model including the management of complexity, because the possible outcome scenarios are too great to effectively plan for. Even in a controlled space with strong defences, a crowd can still present a significant surge to any area that may be perceived as a prime vantage point and could cause crowd pressure to escalate to a dangerous level. Fruin applied this work to crowd crushes at concerts such as the tragedy at a Who concert in Cincinnati in 1979.

Crowd crushes have been avoided in the past by a number of 'defences' including event programming, audience self-removal and security restrictions or capacity constraints. With an increased crowd size these risks need to be closely analysed to ensure adequate defences are designed and set to ensure the risk does not become an incident.

Crowd movement risks

The top three crowd movement related risks are:

◆ Crowd crush in a bottleneck from crowds moving to or from performance viewing areas or access or egress, or to seek shelter from inclement weather, storms or hail;

◆ Crowd crush in the key viewing locations due to limited crowd capacity, excess patrons moving in and out of key viewing locations before or after performances / displays or to get food and beverage or toilets;

◆ Emergency scenarios within the key viewing areas or access corridors causing a panic and crowd rush.

When considering mass movement of crowds and human behaviour, the levels of complexity create a multitude of interrelated risks. There are various models that consider risk and incident causation.

Risk versus incident causation

Structured methods include formal security risk assessment. Security risk assessment is outlined in HB167:2006, currently pending revision and previously attached to AS/NZS 4360:2009 now superseded into ISO 31000:2009. It provides a landscape of security risk assessment for people, assets and information. The primary variation of security risk assessment from standard risk assessment is the acknowledgement that people may wilfully or maliciously commit an act which changes the likelihood or consequence probability.

In terms of crowd management, the emotion of the crowd and their desire to respond and move towards the performers or viewing areas is the additional variable element in the assessment process. To account for this possibility a threat and vulnerability assessment is added.

Applied security risk assessment overarching methodology is not used as much as it should be for crowd management due to the lack of experienced practitioners. As an alternative this chapter considers next other methods of reviewing incident causation that may be more simply applied for crowd management.

The Swiss Cheese Model

For crowd management techniques, James Reason's 'Swiss Cheese Model' has some relevant applications. This model was proposed in the 1970s to identify the defences that failed in allowing a major incident to occur. Prof. Reason used the analogy that major incidents occur when the holes in various layers of 'Swiss Cheese' lined up, therefore allowing a hazard to become an incident or loss. He proposed that the hole in a layer was like a failure and when failures lined up, and all the layers contributed to the process, the hazard became an incident.

The model proposes a given system is broken down into defensive layers or slices of cheese. These slices are the defensive elements that exist between the decision making process and failure causation. Holes exist in each layer, representing opportunities for an accident or error to occur, allowing a clear passage through to the subsequent layer. Each slice is dynamic in nature and holes are continually appearing, disappearing and changing position. When holes align through all layers an accident or error is able to occur.

To develop layers for crowd management planning the initial consideration would be to consider the factors that affect crowd management. The fundamental factors would follow the FIST model described above.

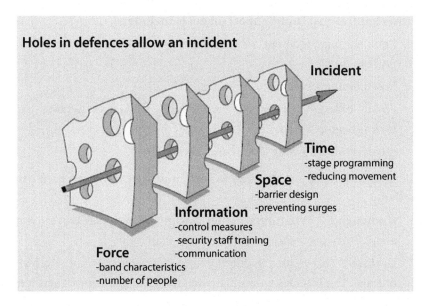

Figure 6.2: Adapted Prof James Reasons' schematic, incorporating Fruin's FIST dimensions.

The factors are:

1 **Force** and emotion relate to how excited the crowd is. This includes their anticipation to get into an event, the size of the crowd, their intention or desire to get to a certain place, such as the barricade, preferred viewing location and need to rendezvous with family or friends

2 **Information** relates to the expectations and information given to the crowd relating to entry procedures, event timings, information related areas that may have imposed entry restrictions due to Conditions of Entry.

3 **Space** relates to how much physical space was calculated for the people and time relates to if the audience is trying to move in a calm and structured manner or in an excited state, i.e. when performances start, or a panicked/frightened state if a live threat emergency exists.

4 **Time** is time between acts, act delays, amenities queues, food and beverage delays and any other instance that can compound scheduling delays.

Two types of holes can appear in defensive layers, the first type is known as active failures caused by a human interaction.

Active failures that could impact on goals include;

♦ Poor site management such as leaving or allowing food stalls in pathways,

♦ Poor communication and briefing,

♦ Failure of security to implement and hold crowd flows when capacity has been reached,

♦ Incorrect decisions by Command Centre.

The second hole is a systematic failure in design. This can be attributed to:

♦ **Knowledge** – What information was given to the crowd, such as about entry procedures and amenities availability ?

♦ **Culture** – This relates to emotion and characteristics of the crowd, but also the interaction and characteristics of the control services, such as security and emergency services. What has been the crowd perception of security control in the past?

♦ **Training** – relates to the understanding, experience and training procedures by the operators or the controllers of the event.

♦ **Design / engineering** – relates to types of barriers used and how they are installed. Will high pressure points have mojo barrier, 'mega gate' and 'gate keeper' systems rather than bike rack and fence panels.

♦ **Procedures** – relate to how crowd pressure is reduced, released, or how people are removed for their own safety out of a tight crowd situation. A special security team needs to be briefed trained and prepared.

The HAZOP study flow-on consequences

A hazard and operability study (HAZOP) is a structured and systematic examination of a planned or existing process or operation in order to identify and evaluate problems that may represent risks to personnel or equipment, or prevent efficient operation. Originally designed by ICI chemical company in the 1960s it survives today as a risk assessment and qualitative tool to test the design intent.

Essentially this was a risk identification process, first, to ask the right questions to start the thought process. The HAZOP technique was initially developed to analyse chemical process systems, but has later been extended to other types of systems and also to complex operations and to software systems.

This is easily adaptable to crowd management because the design objective of crowd flow, crowd management and barricade design is often tested once a sedate crowd becomes excited or forceful or too many people try to approach too small an opening, therefore building up crowd pressure, which results in the hazard of crowd crush.

HAZOP requires the design objectives to be tested by a structured scenario called *deviations* to consider the consequences and the flow-on consequences. Deviations for a crowd management HAZOP would include:

◆ External environmental factors such as the crowd approaching too fast, creating a build-up of pressure, or the crowd being excited or forceful in anticipation of the event.

◆ Internal influences, mood or mood-influencers like alcohol. The crowd becoming forceful or violent due to concerns of inequality or disappointment due to delayed act or cancelled act.

◆ External influences, which may affect rational judgment and exaggerate human panic in a crowd crush. Police security response.

◆ If the pathway of the crowd is convoluted and or constricted at peak load and flow times (including 2-way traffic), crowd flow may slow or cease, which in turn allows a build-up of pressure to occur, which may end up in a crush situation if no other controls are in place.

The design objectives are to ensure that no matter what the crowd flow, a crowd crush does not occur. In engineering sense this means a shut off valve must be installed wherever pressure can build up and there is a system to monitor and activate that valve when required.

Whilst all planning procedures have some benefit, the majority available generally consider administrative tasks and communications flows from a command and control perspective. In line with the systems thinking emergent behaviour characteristics and aligned with the crowd management proposals suggested planning methods based on complexity, early detection and fast response appears to be a strong support to the administrative plans.

Planning considerations for an emergent crowd includes identifying the characteristics of the agents and actors. The agents are members of the crowd, their motivation, emotion, focus and energy. All of these agents need to be considered and taken into account. The corresponding actors in the system are the regulatory stakeholders, including police, security, government (and government legal frameworks), and land-owners or venue operators.

In a complex environment, observation of the emergent patterns is more important than an inflexible pre-planned approach. As crowds can be as diverse as humans themselves the full spectrum of behaviour and interactions has to be considered. Practically this means that although pre-planning is always completed and used in preparation and communication for actors and agents once the event is underway, and crowds are moving, management comes down to observation, assessment and response. Snowden and Boon (2007) suggest that within complex adaptive environments the leader's job is to probe, sense and respond. Hence crowd management needs to be led "from the ground". The most experienced person needs to be able to probe areas of concern, sense emerging issues such as growing volumes and increased densities and respond appropriately, such as by calming, diverting or detouring crowds. This is at odds with some major events where the leaders are requested to be in a central control room, it is not practical to apply the same probe sense, respond when not connected with the crowd.

The Green Guide (2018) capacity estimations and variations of DIM-ICE (Design, Information and Management for Ingress, Circulation and Egress) all lack the real-time ability to respond to changes in crowd .

This style of management is supported by military tactics, where once the battle has started the Field Commander has the ultimate call as the assessment and judgement from the ground. In a crowd management context the same applies as management from the field more accurately allows critical crowd metrics such as density, flow and mood to be observed. Observations from a control room then serve as a supporting function.

Crowd practitioners intuitively observe and react to the relationship between crowd density (the number of people per square metre) and the crowd flow speed (i.e. the number of people moving through a 1 metre gate per minute) and the corresponding effect that has on the crowd mood. Crowd mood is the emotion which generates concern, anxiety and fear or anger and if it continues to escalate can turn to panic. Once a crowd starts to panic, injuries and chaos can occur.

Practical experience has shown as crowd density increases above 2 to 2.5 people per square metre and as crowd flow decreases below 30 people per minute per metre, mood starts to deteriorate. This tipping point can be predicted if accurate quantitative data can be gathered accurately. Actual applications show that when moving in a platoon of people, a mass or large sausage forms once crowd densities approach the 3 people per sqm measurement, and people become concerned about their freedom to change direction and speed, and on the verge of being reduced to a shuffle that concern starts to escalate.

Traditional documents such as the Green Guide suggest emergency egress is calculated at 82 persons per metre per minute on a flat surface. However this is based on free-flowing passage to an empty space and not within a crowd such as at the exit door. Physically up to 82 people per metre per minute can exit through a door but actual observations do not reflect consistent evidence of these egress numbers or evidence of a stable linear approach to an evacuation enabling the theoretical numbers.

When there are crowds in front of queues, once the walking speed is reduced to around 30-40 people per metre per minute human concern also begins to escalate.

Emergent crowd management techniques

Real-time metrics

Complex adaptive systems are defined by their connectivity. Humanity is a complex system. Crowds are a concentrated example of this. Crowd management ecosystems develop very fast in relation to other social ecosystems and hence any actions and reactions need to match the speed a crowd can grow and change. The larger the crowd, the higher potential in crowd energy is possible, the faster the response needs to be. In crowd management, minutes can cause significant changes, therefore all possibilities and scenarios need to be considered.

Crowd management practitioners intuitively observe 3 main crowd metrics.

1   Crowd density, how many people per sqm metre.

2   Crowd flow rates

3   Crowd mood.

Density is a better calculation of than a spatial capacity because it can be assessed for any area, as people will move into space or be pushed into space depending on force and pressure from behind.

New techniques are being developed to monitor these real-time metrics which provide better insight to crowd movements, mood and possible reactions based on considering crowds as part of a complex adaptive environment displaying emergent behaviour.

New techniques involve the observation of density, flow and mood, which are recorded and the data automatically analysed. The analysis applies an algorithm which will show when the crowd mood begins to fall. Modern data science uses data to self-signify the equation rather than trying to formulate an equation to signify an answer. These techniques are the new paradigm of planned reactive crowd management to mitigate risk of emergent behaviour. Cameras are now being programmed to count crowds and use facial recognition technics. This also provides the basis for crowd density and crowd flow observations to be observed and recorded. The advancement of data analysis technics and the large data sets (millions of observations) by computer programed cameras has produced significant and reliable predictive capability. Suggested frameworks for decision making with crowds that use Snowden's 'sense analyse respond' and 'probe sense and respond' require a real-time reaction. Real-time response requires informed and educated observation, correct and unbiased interpretation, and analysis to determine when an actionable response is required. Modern techniques have been developed to collect crowd data exceeding all traditional manual observations. New technology has provided techniques to analyse video images to count crowds, and calculate density and crowd flow rates. These techniques where not accurate when low data sets were made from manual human observations but with cameras now recording hundreds of thousands of data points per hour, this data is becoming very accurate and reliable.

Prior to big data analysis it was assumed that the relationship between density, flow and mood was consistent. That is, as crowd density increased and crowd flow decreases, mood degenerated from neutral, to anxious, to concerned, to fear.

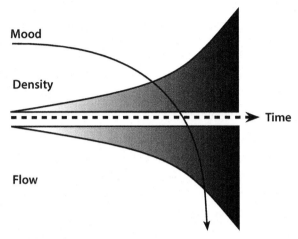

Figure 6.3: Symbolic visual representation of the relationship between crowd density, flow and mood.

This mood continuum is significant to crowd managers as it means the internal changes in patrons' mood will affect the system and outputs, as well as result in crowd dissatisfaction. Once data began being recorded and analysed it became apparent that (in line with complex adaptive environment) the relationships were infinitely variable and so the fixed algorithm of relationship assumptions was abandoned. The new data science approach allows a self-signification of the data. This means there are so many variables that affect density and flow, such as type of crowd, demographics of crowd, spatial areas, and external factors, that a fixed point cannot be determined. Real-time monitoring allows the data points to show when the factors will affect mood. The case study below is an example of the developments in this area of prediction.

Case study: Dynamic crowd measurement

Data patterns have shown that the fall in mood can be predicted by 2 to 6 minutes once changes in crowd density and crowd flow occur. This is very helpful. It means that at any area of concern, front gates, transport hubs, internal bottleneck or gateways, camera vision can record data on the build-up of a crowd to the point when mood is affected. In fact the short warning that can be provided by real-time data collection provides time for pre-arranged and pre-set resources and staffing to implement crowd flow calming or reduction techniques, such as flippers to divert crowds away from a hot spot.

The principles of dynamic crowd measurement were applied to a major event. The event involved a dynamic pedestrian crowd, who meander, stall and move in a free-flowing manner through a complex urban environment. As it takes place in a city street and a precinct environment is not possible to design-out crowd movement limitations, e.g. restrictions and bottlenecks. As a free, open event with a long duration, attendance patterns are unpredictable, crowds can arrive and accumulate at will throughout the event operational period. Held over more than 20 consecutive days it presents significant crowd management challenges, with multiple scenarios that can occur in any sequence and at any time.

A mechanism to enable real-time adaptive response is essential.

To facilitate a dynamic response to emergent scenarios, predictive data analysis is applied. Crowd observations are recorded as measurements by experienced crowd controllers on an App, including crowd density, flow and the mood. A computer software program processes this information to raise alerts when there was a significant increase in density or a significant decrease in crowd flow and/or any sequential change in mood.

These observations, collectively known as dynamic crowd measurement (DCM), allow real-time calculation of crowd metrics to assist with responses.

In alignment with the complex adaptive environment management strategy, the observations showed patterns forming at certain pinch-points or bottlenecks, and as the alerts were received crowd pressure relieving strategies were able to be put in place. It has been demonstrated that by observing these metrics and detecting early and weak signals which would have affected crowd mood within the crowd mass in real-time, 3 to 6 minutes advance warnings of an impending crowd management issue (crush or crowd congestion) were realised.

Real-time crowd management does not absolve the need for pre-planning, including the traditional spatial capacity and DIM-ICE models. But effective real-time crowd management also requires resources, staffing and infrastructure (barriers) to be allocated to various staging areas so they can be quickly implemented to mitigate risks by reduction in crowd flows and pressures.

The next level of data collection is to automate, via the use of smart CCTV cameras, using algorithms which records density, flow and mood in crowds every second. This software is intuitive, and uses machine learning techniques linked with convolutional neural networking to identify density, flow and mood with usable accuracy. Dynamic crowd management has transitioned from emergent theory to active crowd management strategy.

Crowd management in a complex adaptive system (CAS) starts with a boundary constraint, such as no aggression, against others, stay with laneways / queues or move in one way direction. Then it introduces a catalytic probe, known as an attractor, that will create a pattern, such as barriers and check points to create a one-way flow.

This method allows real-time management of the crowd to maintain mood and therefore maintain behavioural outcomes. Dynamic crowd management does not substitute for effective pre-planning, which remains essential, as do two of more of the other assessments techniques detailed above.

The three variables that can be managed in a CAS are the boundary conditions, attractors and the amplification of the attractors. Crowds are an ecological system not an engineering system. A boundary condition for a crowd would be the local laws and precedents, such as no open aggression or civil unrest. An attractor in the crowd situation could be a set of barriers to create a one-way system, this creates a pattern to enable the implementation of more attractors. This is the extent of available manipulation and management of a complex adaptive system.

Review checklist of planning methodology

These two chapters have provided several approaches to crowd management planning. The evolution of this theory has been diverse, varied and contested. Most practices have some application, yet no single application has complete coverage requiring a combined and staged approach to crowd management analysis.

All of them are relevant and applicable, but a single method of planning is not sufficient. They should be used in combination, so several methods can be used in a holistic approach

Pre planning

1   Spatial capacity emergency egress planning , safety planning.

2   Information gathering DIM-ICE RAMP analysis.

3   Risk management based research risk identification , consultation cycles.

4   Crowd characteristics, FIST model and Swiss cheese layers of defences.

5   HAZOP vulnerability considerations. Too much , too fast , too emotional?

6   Real-time management measures, complex adaptive systems, boundaries/ attractors and amplification methods. Data gathering.

7   Learning and continual improvement.

Section Three
Security

Travis Semmens has been hands-on in the centre of crowd management and security field for over twenty years. He is the Managing Director at Australian Concert & Entertainment Security Pty Ltd.

Travis is the person on-the-ground who the police and event managers always consult for timely and vital decisions. His work includes crowd planning and management for events of over a million people in complex urban environments. His work includes liaising, meeting with, managing and consulting to private security guards, crowd management experts and staff and the numerous levels of local, state and national police, military and emergency services.

Travis and his team have been training event security staff for many years.

Travis was assisted in this section by **Aaron Tran**, an expert on security and founder of Vardogyir, a company at the forefront of using machine learning and data analysis for security threat analysis.

7 Security Theory: Process, definitions, tools and techniques

Introduction

The widely respected Abraham Maslow's *Hierarchy of Needs* describes our basic requirement for safety and security as just above food, water, warmth, and rest. For the purposes of this chapter, safety is considered in the context of event security. In this context safety is an emotion that is affected by the trust a patron places on the signals, signs and feelings they detect when they review a place, an event, buy a ticket to an event, or attend an event. In essence, the relationship between a patron and a security provider is one of trust. Patrons attending events have an emotional investment in an event, based on both their expectation of the event itself and on their awareness of the risks of attending an event, formed via knowledge of security incidents at other venues around the world. The security profession, on the other hand, invests in the event process through planning, implementation, and application that needs to be robust and stable to fulfil patrons' trust and maximise their return on investment, and to prevent failure or any other incident that may significantly damage the event.

This chapter will explore what security does to make people feel safe and to prevent the loss of assets. It will use a systems theory approach to discuss the interrelation and interaction of the various dynamic aspects of the different parts of the security process.

What is security?

Security is the planning process that designs and creates the framework for supporting the trust placed by the public in those managing their safety at an event. It also protects the organisers of an event, and the assets invested in making that event a success. Successful security planning is considered a basic requirement for a successful event. This chapter details the application of methodology and processes to ensure an event both feels safe and is safe.

There are three key categories to consider when ensuring successful security. Decisions based on consultation and research need to be made in relation to:

♦ people,

♦ assets, and

♦ information.

The *people* category includes safety and security of patrons, employees, suppliers, and other stakeholders. *Assets* include infrastructure and intangibles like goodwill, brand names, and intellectual property. *Information* includes websites, databases, operational plans, and continuity and recovery plans. These elements combine to make people feel safe, to inhibit intentional malice, and to reduce petty crime and criminal behaviour.

These elements of security will be detailed in the sections of this chapter, as follows: what is security; security planning processes and methodology; emergency management; assessment of security threats; and planning inputs.

Methods to ensure security control can be broken down into the phases:

♦ **Preventative** — reducing the likelihood of risk, for example, Physical layering and separation using stand-off areas and barriers, closed-circuit TV (CCTV), entry searches; venue and service level lock downs and control.

♦ **Detective** — seeks vulnerabilities and gaps for correction, for example, audit reviews, penetration testing, information gathering, intelligence sharing with authorities; white level inspections

♦ **Corrective** — reducing the severity of the consequences after an incident, for example, medical first response, preparedness training and readiness with counterterrorism police and army. Rapid security deployment.

A common approach to security planning and management is to engage a security consultant or planner and then a contractor to deliver the services, so the key elements outlined above can be incorporated and independently assessed. This is likely to involve the event organisers setting security goals for the event and then auditing of delivery goals against a security plan by the consultant/contractor. As security planning needs a high level of independence, the advantage of engaging a consultant and/or contractor is that they provide outside eyes and views free from bias.

Systems

Given the dynamic nature of events and crowds, bringing security goals and actions together in a way that is holistic and integrated requires a systems approach. A system is an integrated collection of parts. Each part, although it is bounded, is interrelated to other parts. Systems theory seeks to understand this and to predict how developments and changes will affect the system as a whole.

General systems theory is about broadly applicable concepts and principles, as opposed to concepts and principles applicable to one domain of knowledge. It distinguishes dynamic or active systems from static or passive ones. Active systems are activity structures or components that interact in behaviours and processes. Passive systems are structures and components that are being processed.

Security implementation is an active system because changes made to increase security presence and security controls and response will change the other actors and agents in the event environment. This may be a positive change, for example, increasing the number of guards can make a crowd feel safer or happier. On the other hand, increasing the guards may increase the crowd hostility.

When developing security planning and methodology, a systems approach is the overarching theoretical framework on which the process relies.

Security planning processes and methodology

Consistent with the systems theory approach outlined above, a security team providing service at crowded places and events may be like a machine. Goals and objectives are defined, every job detail is specified, and all activities are planned, organised, and controlled. If there are weaknesses in this machine like structure, then adapting to changing circumstances will be difficult because it is designed to achieve predetermined goals.

A simplified model of security planning is shown in the flow chart in Figure 7.1, *Security planning process flowchart*.

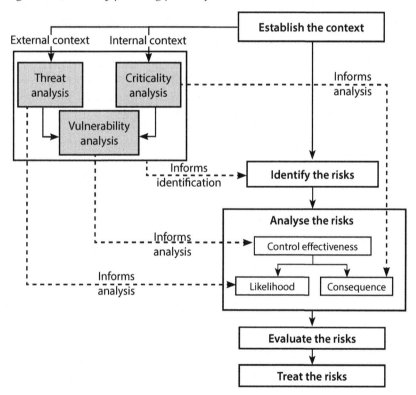

Figure 7.1: Security planning process flowchart. Source: Australia/NZ standard HB167:2006 Security risk management.

The first step is to establish the context of the security risk. Is it external or internal? From this it is possible to determine threats and vulnerabilities. Note the grey sections highlight where planning has to be robust enough to create resilience to overcome targeted and calculated attacks.

These grey sections point to the factors a criminal perpetrator will use to maximise damage. This includes looking for critical infrastructure to make the attack successful as well as the attractiveness or vulnerability of an attack. Security planning can ensure the design does not have any attractive assets or achievable asset for a determined infiltrator via a number of methods. This can be considered as identifying risks, which are analysed to create a plan that both manages security in the lead-up to an event and develops a response to potential risks unfolding during an event.

If security is inadequately planned or the response to security incidents is not adequate, authorities may refuse permission to continue an event or the patrons will not return to the event due to safety concerns. This is particularly the case in the current climate in which modern public events require a security presence to assure the crowd their protection is considered and to allay fears. Terrorism, random shooters, and vehicle-borne aggression have caused anxiety and reduced the number of people willing to attend public gatherings.

Like the machine metaphor used above, security planning is an ordered, process-driven task. The inputs outlined above are used to develop and implement rules and control methods. These will all be mapped out in a security planning document, which is a thorough guide to all foreseeable aspects of event management.

The plan incorporates the following strategies:

1   Physical security recommendations (access control infrastructure, crowd management infrastructure communications systems);

2   Administrative security controls (accreditation systems, preventative security procedures, incident response procedures);

3   Human resource (manpower) security controls / deployment (control and command, static and response).

It relies on coordinated application of effective risk management to mitigate security and safety risk hazards, including:

◆   Event overlay considering safety by design principles;

◆   Effective access control infrastructure;

◆   Effective access control accreditation systems;

◆   Effective access control policies and procedures;

◆   Committed harm minimisation practice;

◆   Effective communications systems and protocols.

Prior to creating the event plan, the vulnerability (attractiveness of assets for loss) and level of protection (risk appetite) the organisers are prepared to pay has to be established through a scoping instruction. Research, internally within the event infrastructure and externally through global databases, media platforms, and reviews, needs to be undertaken. In group discussions on how to respond to issues ,the security planner is primarily a facilitator and moderator of all risk information relating to practical prevention.

The event security plan contents page is shown in Figure 7.2, *Security planning document outline*. This can act as a guide to the relevant information needed in the plan.

☐ Table Of Contents
☐ Introduction
☐ Planning Information
☐ Scope Of Document
☐ Correlation With Related Plans   .
☐ Integration With Related Stakeholders
☐ Event Summary
☐ Event Site Profile
☐ Theoretical Capacity Analysis
☐ Capacity Versus Evacuation And Dispersal
☐ Event Demographic Profile
☐ Entertainment Genre
☐ Patron Demographic
☐ Event Operational Profile
☐ Security Crowd Management Strategy
☐ Physical Security Infrastructure
☐ Security Command Locations
☐ Security Sectors
☐ Security Management
☐ Crowd Control
☐ Security Command Organisation [Event Operations]

☐ Communication Radio Network

☐ Loud Hailers

☐ Security Operations Overview

☐ Security Schedule And Ratio

☐ Emergency Management

☐ Security Function Specific Operational Plan

☐ Supplementary Crowd Management And Security Planning Considerations

☐ Asset Register

☐ Crowd Management Event Reporting

☐ Security Procedures

☐ Annexure 1 – Security Schedule

☐ Annexure 2 – Master Site Plan

☐ Annexure 3 – Show Stop Procedure / Re-Start Procedure

☐ Annexure 4 – Expected Ingress Corridors

☐ Annexure 5 – Expected Egress Corridors   .

Figure 7.2: Security planning document outline

This type of security planning document is developed as follows.

1   Data is collected and used as planning information. (Inputs)

2   Event organisers provide a brief in relation to security requirements, from which a profile is developed. (Processes)

3   Stakeholders are identified, such as:

♦   Event organisers

♦   Land owners or venue managers

♦   Local councils and authorities that grant permission for events

♦   Police and emergency services such as ambulance and fire brigade

♦   Artists, whose requirements are usually communicated by an addendum to the artist contract called a "security rider"

♦   Other relevant suppliers, such as barrier and first aid providers.

4   Tasks are defined and focused operating orders provided. These are called SOPs or standard operating orders and are undertaken by security guards. (Processes)

5   The area in which these tasks must be implemented needs to be defined and communicated. (Processes)

6   A chain of command is established with clear instructions on responsibility for routine tasks. (Processes)

7   A communication system must be in place to communicate instructions to attendees and to refer any irregularities back to management to make decisions or allow variations to the routines. (Processes)

From the *Safe and Healthy Crowded Places Handbook* (AIDR, 2018), the key security questions to ask when developing a plan include the following:

◆   How will all security personnel be identified (for example, uniforms, IDs and tabards identifying roles)?

◆   Will their identification be clearly understood by the attendees and staff?

◆   If private security officers are to be used, what will their role and function be? How will their services integrate with the police? Are they permitted to work outside the venue?

◆   What policies will be enforced in relation to minor offences on-site so that discretion is exercised consistently during the event?

◆   Will there be areas on-site for the collection and storage of significant sums of money, and what security will be in place to protect these areas and off-site banking? Are these areas positioned near road access to avoid carriage of large sums of money on foot through attendee areas?

◆   What arrangements have been made for VIPs? (Police should be able to provide appropriate contacts for considerations in this respect.)

◆   What arrangements have been made for the movement of high-profile persons through crowded areas?

◆   What arrangements have been made for lost or stolen property including the securing of any found property?

◆   What arrangements have been made for lost children?

◆   What arrangements have been made for detection of forged credentials?

We now outline in more detail how and what the development of a security planning document might look like. Security planning requires research, documentation of consultation, and proposed defence mechanisms. It needs to be commenced a minimum of 3 to 6 months prior to an event to ensure all stakeholders, police event organisers, venue managers, and asset managers can review and provide inputs. This can be achieved by using a phased operational profile as outlined in Figure 7.3, *Event security phase profile.*

Figure 7.3: Event security phase profile

A common starting point for security planning is collecting data on previous events. All forms of communication about the previous event, including personal stories, suppliers' gossip, social media posts, venue managers' reports, injury and incident claims and workers' reports are valid sources of data to build a better understanding for the proposed event. If the event is new, broader audience attendance estimations and defences need to be considered until some history is created. Marketing targets and demographics and messaging must be considered. Other inputs are some kind of quantitative incident analysis that includes analysis of incidents of criminal or malicious intent, crime, theft, and illegal unpaid entry, crime conducted within the event, like drug dealing or assault, or whether there are any likely issues with vulnerable people attending.

The site and layered security

Planning commences with a wide view. Sometimes referred to as a helicopter view of the site. The purpose is to create layered rings of protection around the event. The outer layer or standoff distance keeps harm away from the event site. The standoff distance is developed by understanding the threats that could occur at the event. This information is acquired through a consultation process with police, law enforcement and authorities for public order in conjunction with the event security consultant. The factors that will affect the threat level will be: the current threat environment, the location, how accessible the event is to perpetrators, how controllable the area is for security agents and the levels of security to act as deterrents.

The next phase considers the intent and capability of threat agents and actors. Have significant attacks taken place in the past?

The aim of event site security provision is to:

♦ Provide a visible security presence;

♦ Maintain access control;

♦ Maintain a professional security image throughout the event operations;

♦ Implement documented security risk controls for identified risks;

♦ Respond to event and associated stakeholders' security concerns.

Critical to event security strategy is effective communication of access control and implementation of entry procedures, primarily to enforce the conditions of entry as imposed by the event organisers. The best deterrent is to have a distinct presence at all patron entry gates, at all areas of perimeter weakness, and at targeted high risk locations so that there is a visual impact of control systems in place as the public access and observe the event site. This will ensure the public's perception of the event that it is a safe site controlled by proper authorities and should also reduce the likelihood of any incidents occurring.

The inner sector, or sensitive area, of the layered security model includes the stage area, back stage and other back of house areas. The control rooms and event control or management areas are part of it. As well there are many of the assets such as the bars and other cash areas. An example of the security plan for this area is show in Figure 7.4: *Detailed plan for layered security.*

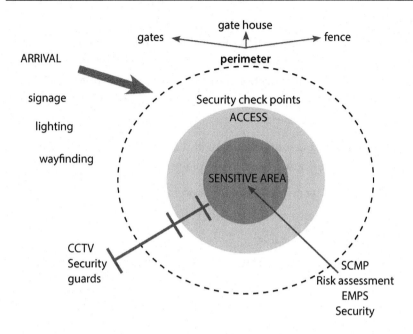

Figure 7.4: Layered security

Attendee's perception of security

This perception is best achieved by having adequate staffing resources and using technology to increase patron accountability. All of which must be agreed on in consultation with the event organisers. The strategy is supported by bright distinct hi-viz event uniforms, with strategic positions supported by hi-viz vests. Event security personnel provide all internal event specific security functions (within the defined event area), hence the planning document is limited to event security provision for internal security services (event asset protection, crowd direction, crowd behaviour, restricted area access control). Any general issues outside the event area are the responsibility of local land authority and/or police (as relevant).

Samples of site security planning by layers

Security duties are then broken down for each area and a command and control information is detailed and provided. *Example 6 Samples of detailed plans for layered security* is from the operational deployment for a concert. This illustrates the layered approach to security planning. More examples of these sectors are found on the accompanying website for this book.

Sector 1 Security Operations: Perimeter and arrival layer

Title:	Security Command
Location:	Refer to site map
Incorporates:	Security management, security radio control, security administration, security event control and local police liaison
Primary hazards:	Communications failure, delayed response to security risks

Manager:	1	Radio Call-sign:	HQ3
Asst Manager:	1	Radio Call-sign:	See below
Security Team #:	6	# Radios:	6
General function:	Manage overall security operations Inform and liaise with all stakeholders as required (promoter, local police, emergency services) Record all radio transmissions Administer overall security operations (including deployment, break relief, equipment distribution, occurrences and incidents)		

Security Positions (# Staff):	Specific role:	Radio call sign:
Security Management / Supervisor (2)	Manage security operations Make decisions on behalf of security contractor Coordinate response to emerging security risks Consult with key stakeholders as required promoter, local police, licensee, medical provider) primarily via Event Control Ensure conformance with plans, policies and procedures Brief and liaise with all security supervisors Ensure appropriate use of resources	HQ3
Administration (1)	Manage and record security sign-on (including any licensing or RSA verification required) Manage provision of licensing and RSA registers Manage and record equipment issue and distribution Ensure appropriate recording of incidents and hazards (incident register, incident cards, incident reports, safety reports, TRAs	Sign On

ECC (1)	Respond to all incoming radio transmissions and distribute or escalate via chain of command	ECC
	Record all radio transmissions.	
	Provide communications function for emergency response.	
	Liaise with local police radio command	
Radio control (1)	Direct all radio transmissions	Control
	Respond to all incoming radio transmissions and distribute or escalate via chain of command	
	Provide communications function for emergency response.	
	Liaise with local police radio command	
Radio scribe (1)	Record all radio transmissions.	
	Complete all duties pertaining to staff member they are relieving	

Figure 7.5a: Samples of detailed plans for layered security, Sector 1

Sector 2 Security Operations: Entry access layer

Title:	Entry
Location:	Refer to site map
Incorporates:	Entry queue areas, all main entry areas, artist entry (excludes internal VIP entry)
Primary hazards:	Access to event of restricted persons (premeditated violence and aggression), restricted and/or or illegal items (drugs, alcohol, weapons)
	Confrontation, violence in entry area (entry refusal related)
	Threats, future claims and prosecution entry search activities
	Regulatory breach (licensing related)

Supervisor:	1	Radio Call sign:	Entry Supervisor
Asst Supervisor:	Appointed from security team	Radio Call sign:	NA
Security Team #:	50	# Radios:	9 + 20
General Function:	Restrict unauthorised access to event (non-accredited persons or non-ticketed patrons)		
	Restrict access of prohibited items in accordance with the Performers National Security Brief		
	Restrict access of alcohol		
	Restrict access of identified risk profile patrons		

Security Positions:	Specific Role:	Radio Call sign:
Entry Supervisor (1)	Brief all security staff in team	Entry Supervisor
	Manage implementation of entry security operations	
	Inform Security Control of all threats, occurrences and incidents	
	Consult with Security Command and co-ordinate agreed response to all threats, occurrences and incidents	
	Manage and co-ordinate all entry security resources	
Queue Management (4)	Provide proactive directional queuing information and instructions to patrons	Queue 1 Queue 2
	Monitor patron behaviour (and request modification where warranted) in queue area	
	Conduct regular checks of the box office	
Bag Search ID Checks / ID Scan Search (44)	Conduct bag search of all bags for prohibited items; Patrons must not, without prior consent of performer's management, bring any of the following items into the venue:	Entry 1 Entry 2 Entry 3 Entry 4 *
	Alcoholic beverages;	
	Cans, metal containers (excluding empty water containers/drink bottles);	
	Illegal or illicit drugs;	
	Skateboards, scooters, roller blades or bicycles;	
	Any structure or item that may be used to erect a structure, or which is capable of supporting the weight of a person including, without limitation, chairs, benches, lounges and stools;	
	containers or other glass objects (excluding prescription or reading eyeglasses, sunglasses and binoculars)	
	Knifes, weapons, packages, gas bottles or fireworks;	
	Ice boxes, drink coolers or eskies;	
	Objects that could distract, hinder or interfere with any performer (e.g. laser pointers);	
	Flags or banners;	
	Whistles, horns, musical instruments, loud hailers, public address systems;	
	Professional digital cameras, electronic or other recording or broadcast devices;	
	Items you intend to distribute, hawk, sell, offer, expose for sale or display for marketing/promotional purposes;	

	...cont. Animals (apart from authorised guide dogs); Any dangerous goods or items deemed by management to be dangerous or capable of causing a public nuisance (e.g. umbrellas). Check presented ID, cross-match with name on ID, supplementary age check on ID (18+), for patrons deemed to be under 25 10 staff to redeploy to response once entry complete.		
Entry Validation Checks (2)	Ensure patrons have correctly passed through the ticketing point.	Check 1 Check 2	

* Note an additional 20 radios will be required for staff redeployments.

Figure 7.5b: Samples of detailed plans for layered security, Sector 2

Sector 3 SECURITY OPERATIONS: Sensitive or inner area

Title:	Stage & artist area
Location:	Refer to site map
Incorporates:	Stage barricade, backstage, dressing rooms, artist catering,
Primary hazards:	Crowd behaviour
	Intoxication
	Unauthorised access production area

Supervisor:	1	Radio Call-sign:	Backstage 1
Asst Supervisor:	NA	Radio Call-sign:	NA
Security Team #:	21	# Radios:	13
General Function:	Maintain access control and restrict access to authorised persons (crew, artists, staff etc).		

Security positions:	Specific role:	Radio Call sign:
Stage Team Leader (1)	Brief all security staff in team Manage implementation of perimeter security operations Inform Security Control of all threats, occurrences & incidents Consult with Security Command and co-ordinate agreed response to all threats, occurrences & incidents Manage & co-ordinate all perimeter security resources Restrict unauthorised access to event	Back-stage 1

Security positions:	Specific role:	Radio Call sign:
Stage Left / Right (2)	Maintain access control to Stage and BOH	Stage left Stage right
Stage Barricade (6)	Maintain access control to Stage	n/a
Artist Area (4)	Restrict access to authorised persons as per pass-sheet Maintain observation of perimeter at all times to restrict unauthorised access. Observation of venue and event assets and reporting and recording any threats or damage to assets.	Artist 1 Artist 2
Artist / Backstage access points (8)	Restrict access to authorised persons as per pass-sheet Observation of patron activities and reporting and recording any irregular activity. Request assistance via Supervisor or Local Police on identification of any suspected threat. Maintain static position at all times. Maintain observation of perimeter at all times to restrict unauthorised access. Observation of venue and event assets and reporting and recording any threats or damage to assets	Back-stage 2 To Back-stage 9

Figure 7.5c: Samples of detailed plans for layered security, Sector 3

Emergency management

A key aspect of developing a security plan is incorporating plans for contingencies to assist with emergency and threatening situations. Security emergencies may stem from severe weather, wind, lightning, storms, physical or structural threats, such as crowd violence, or stage instability. Often the situations, such as a storm approaching, are slow moving and there is time to take a considered approach. However there sudden events, such as a collapse of a stage on the event site that require quick thinking and fast reactions.

Effective emergency management planning requires the following:

♦ Establish a decision-making process;

♦ Focus on opportunities to reduce or manage risk rather than on response to emergencies that may result from risk;

♦   Engage a wide range of individuals and communities;

♦   Promote partnerships and enhancement of relationships;

♦   Foster resource sharing and mutual aid arrangements;

♦   Provide auditable and credible means of reducing risk.

♦   Use language common to decision making in public and private sectors.

Using these guiding principles, an emergency plan should be developed as outlined in Figure 7.6 *Process for developing an emergency management plan.*

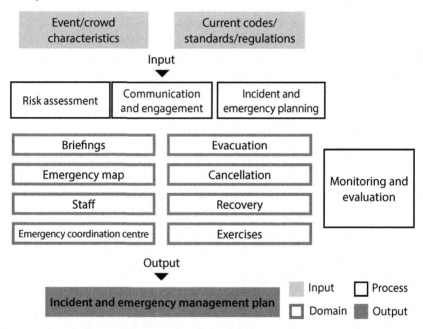

Figure 7.6: Process for developing an emergency management plan. Source: Safe and Healthy Crowded Places Handbook (AIDR, 2018).

Failures

One way of learning about emergency management is to look at the failures and learn from those mistakes. Thinking about the systems theory approach, events are dynamic and if one of the interacting parts fails, then the whole machine is likely to fail. Events are diverse and varied in activities, location, contractors, and management. They are characterised by large amounts of theoretical forward planning followed by an intense period of concentrated activity. The organisers applying emergency management at events often have competing agendas and varying motivation to genuinely adopt emergency management principles rigorously.

Broadly speaking the majority of 'promoters' and 'organisers' of events are blue sky thinkers who attach a very low probability to anything in their event going wrong. In addition, event promoters also dominate the culture of event planning wielding subtle power and influence, which leads to what is known as 'groupthink'. Groupthink, coined by social psychologist Irving Janis, occurs when a group makes faulty decisions because group pressures lead to a deterioration of "mental efficiency, reality testing, and moral judgment". This is why effective emergency planning must be emphasised and why it is important to bring in external expertise.

Failures can be categorised into three types. The first are large events with a historical pattern of success causing organisers to have blind faith in their vision and event planning and execution capabilities. An example is The Love Parade tragedy which claimed 21 lives and injured 510 others. The event was well established, staged regularly, and offered significant tourism benefits to its host city. It used this influence to gain favourable contributions in financial and planning regulatory terms from its host city stakeholders. The fundamental causation for the tragedy was a simple and obvious planning failure – too many people in, and out via the same narrow corridor. The entire crowd (projected at 150,000 people, but actually at least 250,000) were funnelled through a 20m wide tunnel to access the venue. The same tunnel was then planned as a major egress route (and thus, emergency evacuation route). It is apparent that there was no consideration of an evacuation occurring during ingress. The planned tunnel was further narrowed during event installation as a result of fencing and other impediments with a final width of 10.59 m at its narrowest.

The power of influence and the psychology of groupthink, and resetting the contextual framework for emergency planning for large events is the key to improved emergency management practice. A different decision-making environment in which solutions are offered anonymously and subject to blind peer review may be one solution to overcome groupthink.

The second type of failure is that characterised by organisers who pro-actively demonstrate their planning capabilities, a purported culture of compliance, and their intent in order to obtain approval, but who lack the motivation or conviction to execute the required operational details. It is difficult to provide an example in this case as without incident these events are considered successful. Again the groupthink tendencies of current emergency management planning practice are at play here. A possible solution would be the requirement to have independent audi-tors review the implementation of emergency management processes.

The third type is failures in authorities, audits, and approvals. Failures can be systemic or negligent by laziness. An example is Racing the Planet 2011, an ultra-marathon event organised by a Hong Kong based event manager staged in Western Australia's Kimberley region, sponsored by Tourism WA, held on private land in which a bush fire trapped competi-tors and four people were seriously burned and many others trauma-tised. The fundamental cause was found to be the failure of authorities to "meet their own standards" and organisers to consult, even when directed to do so. The event did not involve people with emergency management expertise during the planning process, yet authorities were either made aware of or approved, sponsored, and supported the event without requiring emergency plans to be a part of that approval process.

Similar principles apply to assessment of security threats, which will be discussed next.

Assessment of security threats

Security threat is often assessed subjectively. This is normally done because it is quick and easy and data of incident statistics are often not available. A subjective qualitative scale is shown in Figure 7.7, *Qualitative scale of threat assessment.*

		Intent		
		Little	Expressed	Determined
Capability	**Extensive**	Medium	High	Extreme
	Moderate	Low	Significant	High
	Low	Low	Medium	Significant

Figure 7.7: Qualitative scale of threat assessment

Another method of analysis breaks each security element into the functional areas it supports. Commonly used categories include deter, delay, detect, respond, and recover (DDDRR). A common checklist is detailed in Figure 7.8, *Deter, delay, detect, respond, and recover approach to security risk assessment.*

		Deter	Delay	Detect	Respond	Recover
Physical controls	Signage	Yes	No	No	No	No
	Perimeter barriers	Yes	Yes	No	No	No
	Uniformed security patrols	Yes	Yes	Yes	Yes	No
	Covert security patrols	Partial	Yes	Yes	Yes	No
	Projectile shields	Yes	Delay	Partial	No	No
	Proximity to local traffic (pedestrian and vehicle)	Yes	Partial	Partial	Partial	No
	Open lines of sight (absence of building or terrain cover)	Yes	No	Yes	No	No
	Area lighting conditions	Yes	No	Yes	No	No
	Gating systems	Yes	Yes	Partial	No	No
	Building materials	Partial	Yes	No	No	No
	Vehicle control points	Yes	Yes	Yes	No	No
	Buffer zones	Partial	Yes	No	Yes	No
	Construction codes	Partial	Yes	No	No	No
People controls	Personnel screening	Partial	No	Yes	No	No
	Employee awareness program	No	No	Yes	No	No
	Entry searches	Yes	No	Yes	No	No
	Employee termination procedure	No	No	No	Yes	No
	Staff training	No	Yes	Yes	Yes	Ys
	Personnel movement	Yes	Yes	Yes	Partial	No
	Ethical frameworks & monitoring	Yes	Partial	Yes	No	No

Figure 7.8: Deter, delay, detect, respond, and recover approach to security risk assessment

The reality is all venues, permanent or temporary come with some vulnerability. Even established permanent venues have existing suppliers who have access to the venue and systems that may be penetrated over periods of time prior to the event taking place. Green field sites are even more difficult to secure. Applying risk management principles, the prioritisation of risks is essential. Priorities can be classified into acceptable risks; risks that are reduced to "As low as reasonably practical" (sometimes referred to by the ALARP acronym); and finally unacceptable risks, generally of a higher order of magnitude and higher concern.

Primary considerations in planning is if the event is an unrestricted public gathering or if it is a controlled entry via identification and ticketing. If it is controlled, then planning starts with perimeter security, gate screening, and entry procedures, as well as checking stakeholders and supplier entry credentials. A common trap is to focus on the bigger obvious risks and potential security vulnerabilities while overlooking smaller threats such as internal staff access, suppliers, deliveries, contractors and cleaners.

Permanent venues such as stadia have significant permanent entry security, whereas greenfield sites rely on temporary fencing such as Heras fencing. Discussion of these important inputs into the planning process follows in the next section.

Planning inputs

The inputs required for a security plan include: capacity analysis; site lighting; entry point design; crowd barriers and front of stage barriers for large crowds; security guards; technology; and the media. These will be discussed in turn in the following subsections.

Capacity analysis

Capacity drives all operational considerations in security planning, as the bigger the crowd the bigger the issues may be. In noting specific risks and controls, a security plan should also note details of spatial calculations for venue capacity and the repercussions of exceeding crowd density ratios. Capacity calculation is one of the fundamentals of security planning because of the impact on comfort, space, and organisation, as well as the ability to design emergency exits for safe evacuation.

The global reference for capacity calculations is the UK *Green Guide* (2018). It provides detailed estimates for stadiums and arenas with reserved seating. If the event is ticketed and built in a contained area, the capacity needs to be calculated according to the *Green Guide*. Green field festivals and events require a customised and dedicated approach using the theories of the *Green Guide* but with additional analysis for intra flow bottlenecks and internal crowd movements from stage to stage, or stage to food and beverage areas.

Further information on crowd capacity and measurement is found in Chapter 6, Crowd Planning and Preparation.

Site lighting

The placement of clear white lighting is vital to the security plan. For example, lighting an onsite area will assist:

◆   Access

◆   Flow

◆   Safety issues

◆   Travel paths, and

◆   Enable the attendees to choose their route.

This will be an output of the risk assessment process. For example some events will deter criminals due to the high security presence. Other events may attract them due to the presence of valuables and distractions. There are many stories of theft at stalls in an event occurring during the fireworks. The lights are often turned off, there is a lot of noise and the people are focused on the display. Transition crowd is an attraction for pick pockets. Lighting can minimise these activities.

Entry point design

It starts with layers and standoff areas. Some of the considerations are:

◆   Explosive trace detection (ETD) machines

◆   Hand held metal detectors

◆   Pat downs

◆   Behaviour detection officers

◆   Over camera surveillance

The crowd management plan should ensure the crowd is funnelled so they can be observed. They need to pass through the bollards so they are not open to hostile vehicle attack.

Crowd barriers and front of stage barriers for large crowds

Part of security planning requires the implementation of safety infrastructure. Crowd barriers employed in this manner are "used to relieve and prevent the build-up of audience pressures" (HSE, 1999). The most common type of crowd barrier is the Mojo barrier because of its ability to withstand significant crowd forces. New methods of assessment in the field of crowd management have brought about the introduction of pressure monitoring smart systems. Attached to the base of crowd barriers, Mojo Barriers' *Barrier Load Monitoring System* can provide real time data via computer read-outs of the force being applied.

Splitting the pressure through physically dividing a crowd is a common technique used to make a crowd easier to manage. This divide often occurs in the area immediately in front of stage, where the most enthusiastic and avid fans naturally gather, creating front and rear sections (see Figure 7.9).

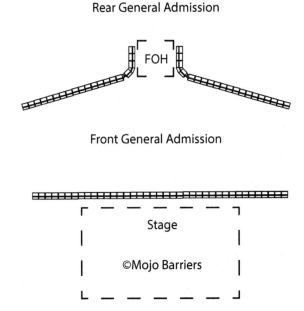

Figure 7.9: Dividing crowd pressure through front of stage barriers

Front of stage barrier systems come in various forms and provide different functions. Selecting the most correct design for an event is dependent on a variety of event characteristics, such as crowd profile, expected capacity, venue dimensions, and the nature of the event (e.g., single concert event or a multi-day festival). For example, trying to restrict sideways crowd movement can be aided by the use of a T-barrier, while preventing crowd surges and crushes from the rear may require a D-barrier. A few examples of currently used designs are found on the website accompanying this book.

The role of event staff and security personnel is paramount to the successful function of barriers. Certain specific guidelines are to be adhered to, such as the ratio of guard per metre of the front barrier and correct response to crowd situations (though this comes down as much to training and preparation as anything else). Spotters may also be positioned in elevated vantage points to monitor and assess crowd safety during an event.

Further safety measures can be taken to minimise the risk of injury and undesirable incidents. Some current practices include restriction on alcohol consumption within the confined area, prohibition of large bags or backpacks, and preventing projectiles being brought into the front barrier. At large scale events, where the crowd stretches back to a distance where sound and visibility become limited, the implementation of video screens and speaker towers become a necessity. This is not only for the enjoyment and involvement of the audience but also to reduce the tendency of patrons to push forward to obtain a better view and clearer sound. At multi-stage events where attendees are likely to be moving between stages all day, there should be alternative viewing areas fitted with large video screens, in case the capacity at one stage is exceeded. This can often be the case for headlining acts. For example, the Rock Werchter festival in Belgium functions with a Main Stage and a Dance Tent. For the majority of the festival the Dance Tent provides ample room for the crowd, however during the headlining set it is almost always at capacity. By providing an additional overflow area complete with a live telecast, VIP and disabled viewing deck, and speaker towers, organisers do not have to put excessive pressure on the tent's barrier system.

In the event of emergency, where evacuation is required, an 8 minute timeframe is the accepted standard (HSE, 1999). This should be facilitated via barrier design. Separate emergency egresses for each section of general admission standing area is needed to avoid congestion at other egress points, as it is well documented that many people will naturally try to exit through the same familiar point through which they entered in an emergency. The incorporation of removable gates or panels within a barrier system to increase egress efforts may also be required. As divided barrier systems allow for greater access to more patrons, it is vital that in an emergency this advantage be used for communicative purposes, providing the crowd with accurate and timely information.

Personnel, including guards and guard training

Effective event management includes having trained and professional personnel. It is important to use guards in a way that is appropriate to the event context. Professors Stott and Reicher (1998a & b) refer to social identity theory to describe the impact of guards in an environment. This theory suggests patrons will evolve themselves to a group or categories to which they aspire given the environmental influences around them. For example, riot behaviour may be precipitated by the arrival of riot police in full armour. Another example, is the use of T-shirt security or peer group security where the crowd aligns and identifies with the security forces, so rock concert security guards with long hair and wearing T-shirt with 'Security' printed on them may get more cooperation and support than formally dressed security, who may be perceived as authoritarian. In a similar vein a business conference or exhibition security should be uniformed in business shirts, ties and jackets to get the best response from the audience. The same theory applies to guards' personal demeanour and communication techniques.

All the best plans and planning can fail if the contracted security forces are not equal to the task. The key to managing complexity and emerging behaviour is innovation and flexibility, combined with situational awareness and rapid, real-time reaction decision making. These can only be achieved if security management is experienced. Similarly, the quality of guards is essential to an integral service. Most regions recognise that security is a position of public trust and have a licensing system or bond system to verify that guards have had a criminal history check and some training. It is advantageous for the event organiser to meet and brief the contracted guards prior to the event, or even prior to the shift starting, to ensure the event message or vision can be communicated. This ensures guards know how the audience is to be treated and how to resolve conflict. These are key elements of basic training for a security guard, which should include:

♦   Integrity and honesty principles;

♦   Entry procedures;

♦   Search procedures;

♦   Detecting suspicious behaviour;

♦   Conflict management;

♦ Prevention of intoxication;

♦ Incident report writing;

♦ First aid training.

When contracting security forces the key essentials are:

1 Experience, knowledge and planning ability and capability;

2 Current licensing and police endorsement, if applicable;

3 Staff resources and availability – this is essential as many contractors will make commitments without having the resources themselves;

4 Training processes and procedures, radio communications and standard operating procedures;

5 References, licences and insurances.

All security services need to be engaged by a legal contract that sets out the quality standards required and the framework for supply and payment. This contract should be created by an experienced legal practitioner engaged by the event to ensure the event is protected from poor practices or under-supply. An independent manager or consultant is required to audit the service roll out by the security contractor.

An example of front of stage barrier management is found on the accompanying website.

Technology

As the principle of security is to deter and detect unauthorised actions, technology is the answer to greater efficiency, better accountability, and evidence-based incident reporting and learning.

Communication is critical. Security teams need to have not only good relations and information flow with police and council authorities, but an effective communication system for the security force via two-way radios or app messaging. Phones alone are not effective as they are one-to-one information transfers, whereas radios allow all staff to listen in on the messages. The planning input needs to include a specific briefing and training for guards, in a form that can be carried and referred to at any stage. Phone messaging and private network messaging can be used for providing instructions to guards, although many still rely on printed instructions.

Other inputs for an effective security service include technology and tools such as metal detectors, bag x-ray machines, and surveillance technology. There are many functions for cameras at events and this is changing rapidly with detection algorithms and other machine learning capabilities to assist security.

The use of drones is also increasing. Drones can be used (away from crowds) to provide high level views of the site such as the entrance and can transmit real time data to the security team. Drones have advantages in emergency situations to provide information about moving crowds and capacity of assembly areas.

The media

Security management and perception is related to the interrelationship between security and communication and media organisations. Strategic communication should be considered pre-event as well as post incident. Security effectiveness is facilitated by:

1   Having an open and known media contact available at all times – this might extend to a media centre at the event;

2   Briefing the communication team on security risks and developing information of security preparedness, as well as response and emergency management if an incident does occur.

3   Consulting with police and authorities on any pre-prepared statements to ensure there is general agreement within the context of the event.

4   Ensuring a credible and articulate spokesperson is available to deliver/discuss statements with the media and conduct desk top exercises with authorities to ensure that the delivery of communication is smooth and credible.

Conclusion

The chapter was a combination of the overview, theory and practicality of security, when it comes to crowds and events. Security concerns the safety of the crowd, the engendering of trust and the protection of assets. The security is a system that can be planned but must also be quick to respond to emergent actions and behaviour. The security of the site is made up of layers. Each area or sector in the layer has its own security responsibilities and tasks. The barriers system is an example of planning safety.

In the next chapter we describe recent security situations that have affected events and crowds around the world.

8 Security: Hostile Attacks

Introduction

Attacks on people at events and crowds in general are found around the world. It has completely changed the security at events, and event planning as a whole. From cement bollards to bag checks, it is an irritant to every event attendee. The extra security cost of events has risen so high that many events have been cancelled. The celebratory element of the event has been diminished.

However it is not the new phenomena that the media seems to assume. Many countries have lived with political/social inspired attacks for years. If a country has hostile neighbours, terror attacks will occur. Regardless of the statistical evidence and the probability of an attack, people are so worried that the sound of a sharp crack in a crowded place can cause panic and a stampede.

This chapter describes some of the attacks on crowds at events, and what the security agencies and governments require from those who manage the crowded spaces. It is well to remember that no matter what is written here or in government recommendations, the attackers are 'free agents' and can adapt their actions to the conditions trying to prevent them.

Low-capability high-impact attacks

Much focus has been placed on terrorism characterised by *high-capability* in terms of substantial planning, resources and coordination to carry out a *high-impact* attack resulting in mass casualties, destruction and disruption, as typified by September 11, the Sri Lanka church and hotel bombings in 2019 and the Paris attacks in 2015. Table 8.1 shows a sample list of these attacks. Improvements in counter-terrorism and intelligence capability and resources have meant that such attacks have become increasingly difficult to successfully carry out. Low-capability attacks do not possess the drawn out and complex pre-attack stage characteristic of high-capability attacks. The latter are inherently more likely to be detected and intercepted.

In recent years, the threat landscape has shifted towards *low-capability high-impact* attacks which require little or no preparation, planning or training and can be launched suddenly and with little to no warning. Low-capability high-impact terrorism is typified by a highly compressed terrorism attack process where pre-attack stages of planning, preparation and training are undertaken in very short amounts of time or removed altogether. An analysis of the terrorist attacks in Western countries in recent years reveals that in terms of number of incidents, solo perpetrator attacks dominate. Many are isolated, ad-hoc attacks undertaken by one person with little training, planning, resources or formal coordination, but ultimately inspired by extremist beliefs.

Whilst solo attacks may dominate in relation to number of incidents, it should be noted that the threat posed by more sophisticated and organised attacks has not diminished. In the case of the December 2015 Paris attacks, highly mobile small teams of one to three perpetrators armed with explosives and firearms, targeted public places and venues with the objective of maximising casualties in coordinated attacks at separate locations. This was again repeated in the attacks on Brussels which involved two coordinated attacks using improvised explosives targeting the Brussels airport and a metro station. Evidence indicating that the two attacks were linked highlights that despite the large number of perpetrators and the amount of preparation and planning involved, they were none the less able to maintain their capability to successfully launch attacks across two different countries.

Table 8.1 summarises a selection of attacks and incidents directly relevant to crowds and events. This chapter will be using the table to illustrate the security methodology for crowded places.

Name	Date	Location	Description	Attack affiliation
Boston marathon bombing	15th April 2013	Boston, USA	Two perpetrators targeting mass gathering event, use of personal explosive device (pressure cooker bombs) followed by shootout with police. Kills 4 including one police officer, injures 280. One perpetrator killed, remaining perpetrator apprehended.	No official affiliation but was inspired by extremist Islamic beliefs.
Paris 2015	13th November 2015	Paris, France	9 perpetrators divided into three teams armed with automatic weapons and suicide bombs undertake six mass shootings and three separate suicide bombings at 5 different sites. Cafes, sports, music concert, queues	Affiliated and organised by the Islamic State.
London Bridge and Market place	3 June 2017	London Bridge UK	3 attacker drove into pedestrians on the bridge, crashed the car and then stabbed people at the nearby market. 8 killed , 48 injured	Islamists inspired by Islamic State
Sri Lanka 2019 Church and Hotel bombing	21 April 2019	Various Sri Lanka	Coordinated suicide terrorist bombings in Colombo and other towns. 259 killed in three churches and hotels.	National Thowheeth Jama'ath, a local militant Islamist group.
Orlando 2016 Nightclub shooting	12th June 2016	Orlando, USA	Perpetrator armed with firearms killed 49 people in a crowded nightclub. Perpetrator was killed by police.	No official affiliation but perpetrator pledged allegiance to the Islamic State

Name	Date	Location	Description	Attack affiliation
Las Vegas 2017 Music Festival attack	1st Oct 2017	Las Vegas USA	Perpetrator killed 51 people and wounded over 400 at the outdoor concert.	No motive attributed. No known affiliation
Nice 2016 Bastille Day	15th July 2016	Nice, France	Perpetrator drove lorry into dense crowd watching fireworks for Bastille day, subsequently opening fire with small arms. At least 84 people killed, total non-fatal injuries in the hundreds.	Claimed by the Islamic State.
NZ 2019 Mosque attack	15th March 2019	Christchurch New Zealand	Perpetrator armed shot. 51 killed, 49 wounded	Anti Islam motive. No known affiliation
Germany 2016 Music Festival bombing	24th July 2016	Ansbach, Germany	Perpetrator detonates suicide bomb at a wine bar and near entrance to outdoors music festival. Perpetrator attempted to enter festival but was denied entry. 15 people injured.	No official affiliation but perpetrator pledged allegiance to the Islamic State.
Ariana Grande concert 2017	May 22 2017	Manchester Arena UK	A lone suicide bomber detonated explosives among teenage fans leaving the concert. 22 people killed, 150 injured	Perpetrator killed. Attack claimed by ISIS

Table 8.1: A Sample of attacks on people at events

Public awareness of terrorism and other attacks at events

Terrorism has been a persistent feature of the media landscape for many decades. It is clear that the general public is now firmly aware and conscious of the threat posed by terrorism.

Whilst the media does not necessarily provide an accurate representation of the threat posed by terrorism, it does influence how the general public may perceive the threat. Terrorist incidents now bring sustained and persistent media attention, with terrorist incidents being given more attention and coverage from more media outlets. Terrorist attacks in the recent past have reinforced this point, with all recent attacks receiving detailed and intense media coverage both nationally and internationally following major incidents.

Terrorist organisations have also developed more sophisticated media management capabilities, including the ability to independently produce sophisticated media releases and distribute them widely using news and social media outlets. As a result of this increased media attention, members of the public in general and other stakeholders will expect event organisers to place specific security plans in place. These trends form key part of the development of an effective risk assessment framework tailored to the current terrorism and attack risk environment.

Background

Attacks have tended to involve one perpetrator acting alone, or one to three perpetrators acting in small groups.

Many attacks on events and crowds typically involved a lone perpetrator, or one to three perpetrators carrying out attacks targeting a single site or as part of a larger coordinated attack. With increased efforts by intelligence agencies and police in monitoring of suspicious activities, it is arguable that operations involving large numbers of perpetrators and which require a greater degree of coordination, planning and communication, have become increasingly difficult and unviable. Single or two perpetrator attacks are inherently more difficult to detect, because such attacks can be undertaken with little or none of the communication that can lead to early detection, and planning is individual rather than coordinated within a larger group. The capability of single or two perpe-trator attacks tends to be limited, particularly when compared to more organised operations, as was the case in the Paris 2015 and Sri Lanka 2019 attacks. However this must be balanced against the fact that the infor-mation required to carry out sophisticated attacks is now more readily available, as demonstrated in the Boston bombings where the perpetra-tors assembled their pressure cooker bombs from online sources. The effectiveness of a single lone perpetrator attack should not be underesti-mated simply because it involves one person.

The terrorist attacks in Paris and Sri Lanka have involved the combina-tion of small teams of one to three people to carry out multiple coordi-nated attacks at separate sites – in this case, the attacks were carried out by three groups of three perpetrators. These attacks present a particular challenge because they tend to be more sophisticated, whilst still main-taining the agility of small one to three person teams, whilst also being more difficult to respond to because they target multiple locations in a short space of time.

The majority of perpetrators had no official affiliation with terrorist groups but were inspired by them or held similar beliefs.

The majority of attacks were not directly affiliated with a terrorist organisation (that is, pre-sanctioned, coordinated, planned or com-manded by a terrorist group or organised terrorist cell). Instead, the majority involved perpetrators who acted on their own initiative and were personally inspired by or held similar beliefs to prominent and well known terrorist groups or organisations, such as Al-Qaida or the Islamic State. This type of perpetrator challenges the conventional images of ter-rorists as operatives who are trained, coordinated and commanded by an organised group, such as the attacks in Mumbai 2008. Instead, these attacks are more locally planned and coordinated.

Whilst lack of official affiliation usually implies lack of formalised or coordinated training and planning, this does not necessarily hinder their capability. Attacks are planned with information each perpetrator gath-ers (which can vary widely) and instruction and training is also read-ily available on the internet (such as material which provides instruc-tion on how to assemble improvised explosive devices). The capability of such solo attacks will ultimately vary with each individual case: the extent of their planning and expertise, the nature of the target and the speed and effectiveness of the response. Though each attack may not be coordinated and officially pre-sanctioned by a terrorist group, these solo attacks have been approved and encouraged by these groups. As these attacks are individually planned, they are intrinsically harder to detect and stop.

The attack in Nice, France highlights how a lone individual with minimal training can still carry out a substantial terrorist attack and inflict substantial casualties.

Firearms feature prominently in the majority of attacks

A substantial number of attacks have involved perpetrators armed with firearms. The firearms used ranged from small capacity semi-automatic pistols, and shotguns to automatic assault rifles. The type of firearm used tended to correlate with the number of deaths, with the deadliest attacks involving automatic assault style weapons as seen in the Mumbai Hotel attacks and the Christchurch mosque attack. Other relevant factors included the crowd density of the target and the speed of the subsequent response from law enforcement.

Experience of mass shootings incidents as a whole reveals that firearms attacks should not be underestimated, even if they are carried out by a single perpetrator: the Las Vagas shooting and the November 2015 Paris attacks further highlight this point, with the vast majority of deaths occurring as a result of the shooting attacks rather than the use of explosives: 89 of the 129 deaths occurring as a result of the mass shooting at the Bataclan theatre during a concert event, in addition to other deaths resulting from the shooting attacks in other areas throughout the city.

Whilst firearms attacks have become more prominent, improvised explosive devices remain a real threat

It is arguable that the increased presence of security and intelligence agencies has meant that the construction of more sophisticated and reliable explosives have become less feasible – ingredients and components that are required are restricted, limited or heavily monitored and any activity may quickly give way to detection. Even so the threat posed by improvised explosive devices (IEDs) should not be underestimated.

Vehicle-borne improvised explosive devices (VIED) can be particularly effective in inflicting casualties and causing widespread damage to surrounding structures. However, due to restrictions on sale of chemicals and other hazardous materials, and the level of expertise and resources required, VIEDs are the most difficult type of IED to assemble or acquire. As such, the threat of a VIED attack as a whole is relatively low. However this does not mean that it is impossible to create a VIED. In 2010, a car was found with an improvised explosive device that included a mixture of approximately 38 litres of petrol, 3 propane tanks, gunpowder, 114 kilograms of fertiliser, 160 firecrackers and various metal drums, casings and wires for shrapnel. The device failed to detonate but given the large amount of fuel and shrapnel material, a successful detonation would have caused many casualties, particularly given the densely crowded nature of where the car bomb was placed. Measures which counter VIED threats are still relevant, particularly in light of the use of vehicles in ramming attacks.

Small, personal improvised explosive devices (PIED) that can be carried by a single person are more common. The effectiveness of these small personal IEDs varies, though they can inflict mass casualties if deployed amongst dense crowds. The Boston Marathon bombings caused 3 deaths and 280 injuries due to the detonation and involved relatively small pressure cooker bombs filled with material designed to act as shrapnel. The pressure cooker bombs were constructed from readily available materials including explosives obtained from firecrackers and fireworks. The November 2015 Paris attacks featured the use of small, personal explosives in the form of suicide bombs carried by at least 7 of the perpetrators. The unpredictable nature of the use of explosive devices was again highlighted by this attack, with reports indicating that the detonation of these IEDs accounted for only a small percentage of the overall deaths in the attack.

The threat of a bombing attack by PIED remains high, as demonstrated by both the Brussels airport and Ataturk airport attacks which involved the detonation of improvised explosive devices carried in luggage at the entrance areas of an airport terminal. Evidence indicates that the two bombs at the Brussels airport were likely coordinated to detonate to maximise casualties, with the first bomb intended to draw panicking crowds towards the second bomb. Use of primary bombs to inflict casualties and cause panic and then a secondary bomb to cause further casualties amongst escaping or evacuating crowds has been a common feature of many IED attacks throughout the world.

Improvised weapons and methods such as vehicle ramming attacks and stabbings have become more prominent and should not be underestimated

Use of improvised weapons, usually everyday objects such as knives, tools such as axes, as well as cars and other vehicles, have become an increasingly prominent feature of terrorist attacks. Due to their accessibility, use of improvised weapons tends to be characteristic of solo low capability attacks.

Whilst the majority of stabbing and vehicle ramming attacks had relatively limited casualties compared to firearm or bombing attacks, they can prove to be highly effective in inflicting mass casualties under certain circumstances and targets including densely crowded areas. The effectiveness of vehicle ramming attacks was demonstrated by the attack in Nice, France where a single perpetrator was able to drive a truck through a dense crowd, killing 84 people and injuring 308 people. It should be noted that the vehicle ramming attack in Nice caused more deaths and injuries in a single incident than in either of the bombing attacks at the Brussels airport and Ataturk airport.

Targeted sites varied though all had at least some functional, political or cultural attribute

Consideration should be given to how insignificant areas can become significant at particular times or due to a particular event taking place. For example, the promenade where the Nice attack took place was made significant due to the Bastille Day holiday, which attracted both large crowds and was also politically and culturally significant. Music festivals, with the mix of youth and alcohol, are targeted due to the ideology of the terrorists. This has been stated many times in the Islamic State publications. Their publications often called for attacks on markets, festivals, concerts or wherever there were crowds.

The November 2015 Paris attacks targeted a mixture of both politically significant sites (the Stade de France, the French national stadium, during a soccer match when the French President was in attendance) and targets of opportunity selected to maximise casualties in what would be regarded as forbidden activities, such as drinking, music and the mixing of genders (restaurants, cafes and the Bataclan theatre).

Active Shooter	Personal IED
Terrorist attack using firearms. Can range from pistols, shotguns to automatic assault rifles. These have caused the greatest number of casualties.	Attacks utilising small IEDs that can be carried on an individual person, such as a suicide vest or bomb in a packpack or other bags/luggage.
Vehicle ramming	**Improvised weapon**
Attack carried out by ramming vehicles, ranging from cars to trucks to bulldozers, into crowds	Attacks carried out using improvised weapons and everyday objects, including knives, axes and machetes.

Figure 8.1: Four dominant terrorist attack scenarios

Public Places	Crowds
Sites targeted were all places which were open to the public, such as museums, theatres, train stations, airports and public streets.	Sites targeted tended to attract large crowds due to their function or nature, such as event venues or mass transit areas.
Events	**Symbolism and Value**
Attacks also targeted special events including sporting events and events held on significant holidays.	Many of the sites had some sort of value, such as political symbolism (headquarters of agency or military memorial), ideological value (bars, concerts, mosques, churches) or some functional value (airports and train stations)

Figure 8.2: Four key characteristics of targeted sites/venues

Context

As with the process for understanding, assessing and identifying risks generally, context must first be established. For terrorism risk, external context is particularly significant because the threat of terrorism to an organisation is driven by broader political, economic, ideological and social forces and factors. As such, unlike other risks such as safety and health risks, which primarily originate from the actions of persons within an organisation (i.e. workers and staff), effective management of security and terrorism risks requires management of risks primarily originating from outside an organisation (the perpetrators).

An understanding of recent terrorist incidents should form the starting point for contextualising terrorism risk. Table 8.1, *A sample list of attacks on people at events*, provides a reference point to summarise the current terrorism risk environment. By looking at past incidents, those responsible for crowded places and events can develop a basic understanding of how attacks have been conducted, what types of sites or venues have been targeted, and the consequences of such attacks. Based on this understanding of recent terrorist attacks, the next step is to identify dominant trends from these recent incidents. Identifying dominant trends and developments should focus on how terrorist attacks have been carried out. This is significant because terrorism risk develops over time, and certain terrorist attack scenarios such as hostage taking have diminished, whilst other types of attacks such as active shooter scenarios have become more prominent.

From this information the event teams can research all the government requirements regarding these types of attacks. The governments of the USA and the UK, for example, have detailed advice available to event and crowd managers, recommendations as well as legal requirements for security against attacks. Also, the venue and other event stakeholders have their policies which, ideally, will fit with the government requirements.

As many attacks are related to politics and ideology, governments have instigated various alert systems. The threat level is an input into the security planning. The type of alerts and how they are communicated varies in different countries. The basics is that government intelligence agencies such as the MI5 in the UK and the National Terrorism Advisory System of the Department of Homeland Security in the USA published their assessment of the current threat levels. Figure 8.3 lists the threat levels from MI5 in the United Kingdom.

Threat levels are designed to give a broad indication of the likelihood of a terrorist attack.

☐ LOW means an attack is unlikely.

☐ MODERATE means an attack is possible, but not likely

☐ SUBSTANTIAL means an attack is a strong possibility

☐ SEVERE means an attack is highly likely

☐ CRITICAL means an attack is expected imminently

https://www.mi5.gov.uk/threat-levels

Figure 8.3: Threat levels from MI5, UK

Understanding the components of terrorism risk

Terrorism risk is comprised of threat, vulnerability and consequence.

♦ **Threat** describes the probability of a terrorist attack occurring, and in this sense, is a consideration which involves the external context of risk management considerations.

♦ **Vulnerability** describes the probability that an attack would succeed in causing an outcome such as casualties and damage.

♦ **Consequence** describes the extent of the damage caused.

An event organisation can have some influence over its vulnerability to terrorism and the consequences of a terrorist attack and as such, vulnerability and consequence involves the internal context of risk management considerations.

Threat	Vulnerability	Consequence
What is the probability of an attack occuring?	If attacked, what is the probability that the attack would be successful in causing damage or casualties?	If attacked, what would be the extent of the damage?

Figure 8.4: Components of attack risk

Prior assessment

There are a number of tools to assess the event or crowded place provided by governments. These can be a formal tool such as *ANZCT Crowded Places Self Assessment Tool*, or a list of question such as *Step One: Identify the threats* in the UK *Counter Terrorism Protective Security Advice for Major Events*. It is essential that the event team establish communication with the various law enforcement bodies.

Two tools used are EVIL DONE and CARVER

EVIL DONE (Exposed, Vital, Iconic, Legitimate, Destructible, Occupied, Near, Easy) is a framework which assesses the site as a target from the perspective of a potential attacker and is a useful tool to provide a metric of vulnerability.

CARVER (Criticality, Accessibility, Recoverability, Effect and Recognisability) is a protocol used by US Special Operations Forces in assessing and targeting an adversary's installations and analysing threat from the perspective of counter-terrorist forces. CARVER is a useful tool to provide a metric for determining likely consequences, losses/damage, and in turn planning the response.

Effective, Appropriate and Proportionate principles

The foundation for effective security risk management is to align the level of security measures with the level of risk to ensure that these measures are effective, appropriate and proportionate (EAP):

1 **Effective:** how likely are the measures to succeed in preventing an attack or mitigating an attack if it occurs?

2 **Appropriate:** do the measures reflect the likelihood of an attack?

3 **Proportionate:** do the measures reflect the scale of the threat or its potential consequences?

Effectiveness is a judgment of the security measure alone – how well it is able to meet a potential threat. This is then balanced against Appropriateness and Proportionality which take into consideration the costs of security measures.

The EAP framework reminds us that highly effective security measures may not always be appropriate or proportionate: mandatory full body searches, for example, can be highly effective but if the threat level is low, may neither be appropriate or proportionate and can begin to alienate visitors and be detrimental to goodwill.

Guarding and patrol measures

The deterrent effect of security measures such as searches on entry, uniformed patrols and overt CCTV is best approached as a balance between being *visible* and *routine* and *unpredictable* (VRUP)

♦ **(V) Visible:** Make it known that robust security measures are in place. The entrance to the site and the perimeter which separates 'inside' from 'outside' will be the most obvious places to make security measures visible and known (perimeter patrols, searches at the entry, wands and walkthrough metal detectors, barriers, walls, CCTV at entry lines, etc.).

♦ **(R) Routine:** Security measures need to be maintained and not just applied at specific and predictable times which allow perpetrators to adapt. Routine ensures that the deterrence effect is maintained and that occupants and patrons are consistently reassured of a security presence once 'inside'.

♦ **(UP) Unpredictable:** This applies to both how the measures are performed (not making guard routes predictable) but also to the measures themselves. Keeping measures routine but unpredictable is the best way to ensure that they maintain their effectiveness as a deterrent.

Other guarding strategies and tactics

Additional security staff assigned to specialised roles such as 'spotters' at entrances and plain clothes patrols can provide an additional layer of security. Plain clothed guards hold a tactical advantage to notice suspicious behaviour which may be concealed in the presence of uniformed guards. Whilst their role will be to blend in with the crowd, their presence can be acknowledged with a simple sign stating that plain clothes security staff patrol the area – this can assist in boosting the overall deterrent effect as well as reassure patrons.

Provision of additional Control Room Operators (CROs) to increase active surveillance capability has also proved successful in monitoring larger crowds during large mass gathering events. Ongoing communication via radio and mobile phone between the plain clothes guards and uniformed staff will contribute to early intervention and engagement with suspicious persons.

Example 7 Case study: layers of security

A security company was engaged to design a guarding and patrol strategy to improve terrorism preparedness for the large event, following an increase in the National Terrorism Alert Level. The event attracts nearly 1 million people over two weeks.

The primary challenge faced was the need to cover a large area with multiple points of entry, as well as accommodate the large flow of people in and out of the event site throughout the during of the event itself. The company designed a multilayered guarding and patrol strategy comprised of security guards deployed in four different teams with specific roles:

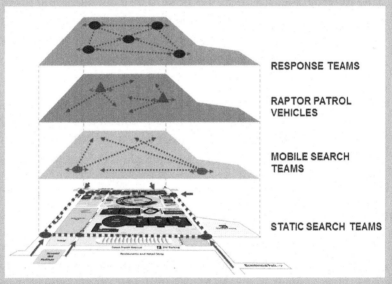

RESPONSE TEAMS

RAPTOR PATROL VEHICLES

MOBILE SEARCH TEAMS

STATIC SEARCH TEAMS

Figure 8.5: Visualisation of the security layers

The **Response Teams** were security guards tasked with acting in an immediate response capacity to provide additional reinforcement and manpower should a specific incident arise.

Raptor Patrol Vehicles were deployed to rove throughout the showgrounds, with the Raptor vehicles (stand-on one person vehicle) further enhancing the high-visibility security presence.

To cover the large area of the grounds as well as the large number of secondary entrances, **Mobile Search Teams** were also deployed to move from section to section and establish a temporary search area. The mobile search teams would conduct searches for a specific amount of time before moving on to another area to conducted searches in another area.

Static Search Teams were also deployed at the five main entrances to the site, with their role being to maintain a consistent presence and search at entry capability.

Layered security

The concept of layered security was introduced in the previous chapter concerning the event site. It is fundamental to the recommendations against hostile attacks as they are so seemingly unpredictable and can change according to perceived conditions. The concept is that if one layer of security fails, the people as still protected. In other words there are multiple redundancies.

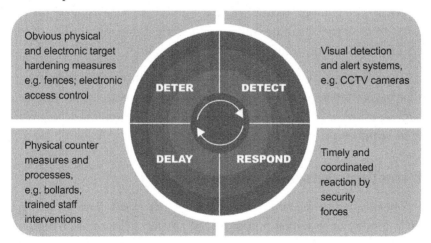

Figure 8.6: Layered Security. From *Australia's Strategy for Protecting Crowded Places from Terrorism*, ANZCTC 2017

"Layered security reduces the likelihood of a successful attack on a crowded place by building multiple layers of redundancy into a site's security architecture. This can be achieved through a 'deter, detect, delay and respond' model. Deterring, also called target hardening, concerns using obvious physical and electronic measures. Detecting includes using visual and auditory alert systems. Delaying means to slow down the intrusion and allow security time to respond. Responding means the use of competent security staff and reliable communications. The actual measures

within each of the levels will differ for each location and will be influenced by a variety of factors, including the purpose of the venue/event, threat advice, history of incidents and existing security measures."

From *Safe and Healthy Crowded Places*, O'Toole, 2018.

Search of persons/bags at entry

Searches form an important part of maintaining the security of a site particularly in context of specific threats (a planted bomb attack) or heightened security risk level (such as a mass gathering/event day).

The establishment of a search zone at the point of entry serves as an additional layer of security which can assist in deterring, detecting and preventing attempted movement of a threat further into the site. For example, search and entry controls at the Stade de France resulted in one of the suicide bombers of the November 2015 Paris attacks being turned away from entry, resulting in him detonating his bomb outside of the stadium to relatively minimal impact and casualties.

A key challenge for the design and implementation of a search of persons/bags strategy is determining to what extent searches of persons/bags be undertaken. It is often impractical for all persons and their bags to be searched prior to a large event.

Detecting, deterring and preventing

Sophisticated and coordinated attacks are the most likely to inflict a high number of casualties and cause the greatest amount of disruption. They will often involve careful planning, gathering of information (hostile reconnaissance) and even rehearsal of the attack itself. The perpetrators of the 2005 London bombings held a trial run nine days prior to the attack itself. Even ad-hoc attacks such as the attack in Nice, France also demonstrated some degree of preparation, with surveillance footage indicating that the perpetrator had visited the site of the attack prior to the attack itself.

Aspects of the pre-attack phase of a terrorist attack can be subtle and difficult to detect. Whilst CCTV surveillance and guarding and patrols by security staff play an important role in detecting, deterring and preventing an attack, even these measures can miss subtle details and changes.

To ensure the most comprehensive level of terrorism preparedness, an organisation must ultimately involve all its staff, and not just designated security staff, into the process of contributing to the overall security of the site/venue.

All staff can play an active role in maintaining awareness and vigilance of their immediate surrounding environment. Indeed, the person best equipped to notice subtle details, abnormalities and unaccounted changes that may be missed by others is the person who consistently works within that immediate environment or space. By developing effective terrorism awareness amongst all staff, and equipping them with the necessary knowledge and skills to identify abnormalities, suspicious behaviour or objects, 'whole-of-organisation' monitoring capability can be achieved whereby all staff collectively contribute to ensuring the security of the site/venue.

White level inspection

A white level inspection is an inspection by staff of the event site and nearby for any articles that are unusual, suspicious or unable to be accounted for. One of the issues with events is the newness of the site. Hence, although a white level inspection is performed by staff who are familiar with the day to day normality of their workplace, the temporary nature of events is a limitation.

Evacuation

The unpredictable nature of terrorist attacks requires sites/venues to develop flexible and adaptable emergency response procedures including contingency planning for emergency evacuation. The attack on the Stade de France in November 2015 highlighted the importance of contingency planning in relation to emergency evacuation procedures, with indications that the attack was intended to cause an evacuation out of the stadium, where subsequent IEDs were then to be detonated. The decision taken by police to facilitate an evacuation onto the field due to the remaining terrorist threat in the immediate area likely prevented further casualties as well as injuries that may have occurred if a simultaneous mass evacuation of all spectators out of the stadium took place.

The terrorist attack at the Stade de France highlights several important points in relation to contingency planning for emergency evacuation for attacks. Emergency evacuation for a terrorist incident may be hindered by confusion or uncertainty in relation to the exact nature of the terrorist threat. For example, during the attack on the Stade de France, it was not until a third explosion was heard that an emergency response was implemented.

Due to this uncertainty, crowd behaviour in an evacuation due to a terrorist attack will not be the same as evacuations for other incidents – the crowd is more likely to be panicked and may rush to points of evacuation. In such scenarios, consideration should be given to a staggered evacuation process involving a safe and secure intermediary evacuation point within the site/venue where subsequent evacuation out of the site/venue can then be facilitated in a controlled manner. For example, evacuation onto the field may serve as a staging ground or intermediary evacuation point, allowing crowds to be moved away from the point of danger but preventing a simultaneous and immediate rush to points of exits.

In situations where an active terrorist threat remains outside of the site/venue, contingency plans need to be developed in relation to keeping patrons and visitors secured within the site/venue itself. Again, identifying possible safe and secure intermediary evacuation points where patrons and visitors may be kept for an extended period of time until evacuation outside of the site/venue is safe and secure should also be considered. Such intermediary points can maintain the safety of the crowd if combined with effective dynamic lockdown procedures which prevents an external threat from entering into the site/venue grounds.

Dynamic lockdown is the ability to quickly restrict access and egress to a site/venue or parts of the site/venue. The overall aim of dynamic lockdown procedures is to restrict both the movement of people towards the threat or dangerous areas, and to slow or hinder the movement of an attacker or prevent them from further access to the site. Where a threat emerges in one area, dynamic lockdown procedures should be immediately engaged alongside emergency evacuation procedures, moving people away from the threat whilst also containing and restricting the forward movement of the threat as well.

Hostile vehicle attack

Vehicle ramming attacks can involve the use of a car, larger vehicle such as a truck or lorry (as was the case in the attack in Nice) or other vehicles such as bulldozers (one such attack using a bulldozer occurred in Jerusalem in 2008). Regardless of the type of vehicle, the underlying aim of such attacks is to ram the vehicle into the target, usually people and large crowds. The attack in Nice demonstrated that such scenarios can result in substantial casualties, particularly when directed against large and dense crowds. Vehicle ramming attacks are a form of sustained terrorist attack – they will continue until the perpetrator or vehicle is incapacitated. If a perpetrator can maintain control of their vehicle, they can inflict substantial casualties against a large and dense crowd as was demonstrated in the attack in Nice. In the case of Nice, the vehicle attack was sustained for a distance of 1.7 kilometres and some 5 minutes before being stopped.

For major venues and stadiums, large and dense crowds congregating around or outside the entrance gates, as well as crowds moving along pedestrian corridors, streets and roads of the site/venue after an event, are particularly vulnerable to vehicle ramming attacks. This risk is further heightened across pedestrian corridors where crowds can only move in one direction (such as a road bridge or corridor which is walled or fenced off on the sides). For such areas, strong vehicle restriction and control measures such as barriers, walls and bollards must be in place to prevent any entry of an unauthorised vehicle.

The ability for vehicles to park near to critical areas should be restricted where possible. Vehicle borne explosive devices are one of most effective terrorist methods due to their ability to hold large quantities of explosives and be delivered very precisely to a target. Vehicle bombs are a high-reward high-risk option for the attackers. They are very effective but require substantial resources and this increases risk of detection and failure. Carrying out such an attack requires substantial expertise and planning.

In light of recent attacks, vehicle controls are now an essential and necessary control for terrorism risk management. Preventing possible vehicle ramming attacks places greater focus on restricting and controlling movement into the site and particularly into crowded areas.

The risk mitigating strategies for vehicle attacks have been the most visible signs of anti-terrorism devices at crowded places. These comprise cement bollards, gates, road closures, trucks used for protective purposes and even Armed Personnel Carriers (APC). The bollards are so arranged to slow down the vehicle. Other impediments can be uses such as landscape features, trees, planter boxes and sculptures reinforced against vehicle impact. These can be placed to create a standoff distance. The UK Counter Terrorism Protective Security Advice for Major Events recommend, for example, 30 metre standoff distance to the event.

Improvised weapon attacks

Improvised weapon attacks typically characterise solo attackers where perpetrators must work with limited resources to carry out an attack. Previous improvised weapon attacks have involved use of knives or other bladed weapons such as axes or machetes. Improvised weapon attacks rarely cause many casualties but the threat they pose should not be underestimated particularly when multiple attackers may be involved.

As the extent of terrorist capabilities using knives or other such weapons is much more limited than an active shooter scenario, consideration should be given to neutralising the attacker. They can be more easily overpowered than is the case in other types of terrorist attack scenarios. Improvised weapons can be used as a backup by the attacker. In the 2017 London Westminster bridge attack the three terrorists jumped from their car and continued their attack on people using knives. This is an example of how the attackers are able to quickly adapt to the situation.

Chemical weapons

A chemical weapon is a device designed to release a chemical. It could be corrosive, flammable or toxic. It strength is influence by the environment such as weather and the size of the enclosed area. The effect on people can range from eye irritation to collapse and death. The recommendations of the government handbooks listed at the end of this chapter are similar to the other risks. Such as:

♦ Pre-plan evacuation routes
♦ Communications

♦   Reporting procedure for suspicious behaviour and objects

♦   White level inspections

♦   Access to first aid

Conclusion

A hostile act in a crowd is a frightening but rare occurrence. The information in this chapter is based on a study of these acts and the recommendations by government and security agencies. As stated in the beginning of the chapter it is dynamic and often unpredictable. Also the attacker can quickly adapt to any security presence. For this reason treat the information contained herein as only a starting point. It is part of the risk management process found in all the other chapters. By its very nature it is uncertain.

Further information

Australian Institute for Disaster Resilience (1999), *Australian Disaster Resilience Manual 12: Safe and Healthy Mass Gatherings*, https://knowledge.aidr.org.au/media/4455/manual-12-safe-and-healthy-mass-gatherings.pdf

Australian Institute for Disaster Resilience (2017), Australian *Disaster Resilience Handbook 3: Managing Exercises*, https://knowledge.aidr.org.au/resources/handbook-3-managing-exercises/

Australian Institute for Disaster Resilience (2013), *Australian Disaster Resilience Handbook 8: Lessons Management*, https://knowledge.aidr.org.au/resources/handbook-8-lessons-management/

Australian Institute for Disaster Resilience (2015), *Australian Disaster Resilience Handbook 10: National Emergency Risk Assessment Guidelines*, https://knowledge.aidr.org.au/resources/handbook-10-national-emergency-risk-assessment-guidelines/

Australian Red Cross and Australian Psychological Society (2013), *Psychological First Aid: An Australian guide to supporting people affected by disaster,* https://www.redcross.org.au/getmedia/23276bd8-a627-48fe-87c2-5bc6b6b61eec/Psychological-First-Aid-An-Australian-Guide.pdf.aspx

Australia-New Zealand Counter-Terrorism Committee (2017), ANZCTC *Active Armed Offender Guidelines for Crowded Places*,

Commonwealth of Australia, https://www.nationalsecurity.gov.au/ Media-and-publications/Publications/Documents/active-armed-offender-guidelines-crowded-places.pdf

Australia-New Zealand Counter-Terrorism Committee (2017), *ANZCTC Chemical Weapon Guidelines for Crowded Places,* Commonwealth of Australia, https://www.nationalsecurity.gov.au/Media-and-publications/Publications/Documents/chemical-weapon-guidelines-crowded-places.pdf

Australia-New Zealand Counter-Terrorism Committee (2017), ANZCTC *Hostile Vehicle Guidelines for Crowded Places: A Guide for owners, operators and designers,* Commonwealth of Australia, https://www.nationalsecurity.gov.au/Media-and-publications/Publications/Documents/hostile-vehicle-guidelines-crowded-places.pdf

Australia-New Zealand Counter-Terrorism Committee (2017), ANZCTC *Improvised Explosive Device (IED) Guidelines for Crowded Places*, Commonwealth of Australia, https://www.nationalsecurity.gov.au/Media-and-publications/Publications/Documents/IED-Guidelines/IED-guidelines-crowded-places.pdf

Australia-New Zealand Counter-Terrorism Committee (2017), *Australia's Strategy for Protecting Crowded Places from Terrorism,* Commonwealth of Australia, https://www.nationalsecurity.gov.au/Media-and-publications/Publications/Documents/Australias-Strategy-Protecting-Crowded-Places-Terrorism.pdf

NACTSO (2017), Crowded Places Guidance, National Counter Terrorism Security Office UK

Section Four
Health

Dr Stephen Luke brings years of frontline experience in crowds around the world to this section. As well as his medical career as a specialist in Intensive Care, he is a Medical Director with Team Rubicon. They fly to disasters anywhere in the world to assist during emergencies. Famed as willing to travel to areas deemed too dangerous by other aid organisations, Dr Luke understands the importance of planning and quick decisions in a complex and time critical situation.

Large gatherings of people bring new health problems as well as logistics issues. It is an excellent example of scaling. It is not a simple linear model. Large events have significantly different health and medical issues to small crowds. In the next two chapters the reader will find the latest information on this perspective of crowd management.

9 Integrating Health

Introduction

Health is inherently complex and negotiating its challenges is the epitome of complexity management.

To the uninitiated, developing an event health plan becomes a crash course in balancing previously unappreciated risk with an ever growing list of needs and cost. All too often this is complicated by the need to negotiate a seemingly endless number of opinions, organizations and personalities, often while learning a new (medical) language.

Managing health in the dynamic and often unpredictable context of crowds is a specialist skill that requires strategic planning and experienced staff, working within effective systems and with appropriate resources. Expenses are real while funds and resources are limited.

Health planning is integral to event management, takes time and needs to commence early. Bringing all parties to the shared realization that everyone fundamentally wants a safe and successful event is an important early milestone.

Event and health managers need to understand complexity management from the other's perspective in order to successfully plan and manage events and crowds. An attempt at translation is provided on the following pages.

Event health planning (Process)

Crowds are inherently risky and unpredictable yet people understandably expect to be safe in public spaces, especially at planned events. As event managers, our goal is to plan and deliver a successful event and health planning often becomes a peripheral concern. As with all other aspects of event planning and crowd management, health planning challenges exist in identifying hazards, mitigating risk and maximising individual and community resilience.

Mass casualty incidents at public events are front page news, especially if terrorism is suspected. Pandemic influenza and Ebola Viral Disease are high profile global health issues. The lethality of heat waves is becoming increasingly understood. Mass drug overdoses at music festivals may result in catastrophic loss of life and inevitably evoke passionate community debate. Why then is health so often an after-thought in event planning?

Health is a complex, insular beast. Health culture and language are foreign, intimidating and often poorly understood by outsiders. Comprising many loosely-bound component agencies and perpetually running at or over capacity, Health typifies many of the challenges of complexity management theory and multi-agency incident command. Despite many similarities, Health is often only peripherally involved in event and emergency planning and regularly engaged superficially and late.

Event health planning is a niche skill and requires significant experience across a range of event types. The use of an established event health planning checklist is strongly recommended but is not a substitute for experienced event health planners. Early and ongoing consultation with local ambulance, hospital and reputable event health service providers is critical.

Within health, each organization involved may have different (and potentially conflicting) priorities. While triggers, risks and outcomes vary, many mitigation strategies share common principles. It is critical that the chosen service delivery model be agreed upon by all interested parties, and be flexible, scalable, adequately resourced and led by experienced health commanders.

Strategic health planning can dramatically reduce the impact of an event and/or crowd on local communities and health services. Consider the following scenario:

Example 9.1: Effect of a large music festival

A large music festival held near a small country town will congest local roads, busy local ambulances and likely exceed local hospital capacity. Without additional on-site health services to reduce this impact and with prolonged transport times to larger hospitals, the availability of road and air ambulances to respond in the local community will be reduced. Delays responding to medical emergencies like heart attack, stroke and cardiac arrest result in prolonged suffering, long term disability or death.

Event managers have a moral obligation, and often legal requirement, to minimize the adverse impact of events on local communities. Local standards and legislation vary widely with a number of excellent resources readily available.

Strategic event health planning can minimize the impact of events and incidents on local communities and health service. Committing event resources to the provision of high quality on-site health services and augmenting local health and ambulance services achieves:

1   **On-site resuscitation and advanced life support teams** – some medical and trauma presentations require urgent and life-saving interventions. Ambulance transport times from events and through crowds to hospital are always longer than the theoretical transport times on a map.

2   **On-site patient assessment, stabilization and management capacity** – allows for patients to be assessed, initial treatment commenced and ongoing management continued, awaiting transfer to hospital. This often results in reduced acuity and can safely avoid the need for an ambulance transport.

3   **On-site primary care and definitive management health services** – allows for minor ailments and injuries to be assessed and patrons returned to an event or discharged home, avoiding the need for an ambulance transport and hospital presentation. Patron satisfaction

is greatly increased when they can receive the treatment that they need while avoiding leaving an event that they have paid to attend.

4   **Health promotion** – increasing resilience through focused pre-event publicity and general community education is an important and effective method for changing behaviour, influencing culture and reducing presentations for medical assistance

5   **Exercising staff, equipment and systems** – mass gatherings provide excellent opportunities to train staff, test equipment and refine systems. Ambulance and health services may elect to make the most of these opportunities

The development and implementation of an integrated health plan is a critical output of the planning process. This health plan must complement, and most importantly not contradict, other event and local health service plans. Furthermore, events provide an excellent opportunity to exercise existing disaster response plans, equipment and staff, especially given the inherent risk of rapid and unpredictable escalation.

Example 9.2: The Boston bombing

The bombing of the 2013 Boston Marathon drew worldwide attention to the vulnerability of mass gatherings to deliberate acts of violence. While many were severely injured, the number of deaths would have been much high in the absence of established and experienced on-site medical first responders.

The report, "Why was Boston strong? Lessons from the Boston Marathon Bombing", highlights the benefits and importance of established command systems, relationships and communication. The use of mass gatherings to exercise and refine disaster response plans is emphasized.

https://www.hks.harvard.edu/sites/default/files/centers/rappaport/files/BostonStrong_final.pdf

Health risk assessment (Input)

Performing a health risk assessment is critical to identify specific health risks and to identify other event and crowd related risks that will impact on other aspects of the event and emergency management plans. This process must begin early and continue throughout the planning process.

Factor analysis of previous mass gatherings (and common sense!) identifies that some events and crowd states carry intrinsically higher, and sometimes highly specific, risks. The World Health Organization list of event characteristics that affect the health and well being of a crowd provides a useful starting point for the event manager.

No tool exists to accurately identify *and* plan for all contingencies, especially given the complex and dynamic nature of events and crowds. The use of a standardized hazard identification tool will assist with the identification and documentation of risks.

Changing a single factor can have dramatic implications for health planning and response. Consider the impact of the changes in Example 9.3, isolated and combined, to a motocross event.

Specific consideration should be given to the health implications of environmental factors. The event management team should monitor weather throughout the event, but this extends beyond exposure to heat and cold extremes, and includes:

♦ The need for adequate shelter from temperature, wind, sun, rain and snow

♦ Triggering of medical conditions (dust and respiratory illnesses)

♦ Seasonal and geographical susceptibility to natural hazards (storms, bushfire, flood)

♦ Disease transmission through contaminated water, sewage, aero-solized particles in confined areas and direct physical contact

♦ Sites at high risk of crowd convergence with risk of crush injury

♦ Locations vulnerable to physical attack, including motor vehicle collisions

Example 9.3: Change from a small event to a larger event

From	To
50 competitors	300 competitors
A local club event	A national championship
Adult competitors	Child competitors
Closest hospital 5km away	Closest hospital 50km away
Closest hospital major trauma centre	Closest hospital small rural hospital

☐ A larger event with more people will have more presentations for medical assistance.

☐ Increasing competitiveness will often increase risk taking and result in more serious injuries.

☐ Child injury patterns are different to adults and not all first responders and medical providers are comfortable managing critically injured children.

☐ The distance to hospital directly impacts on ambulance transport times and lengthens time before that ambulance can attend to other patients.

☐ Small rural hospitals will often have limited staff, resources and experience to manage multiple patients and major trauma.

The impact of each of these (and many other!) factors must be taken into account when considering on-site and emergency patient transport services.

It is crucial that all health service providers are appropriately resourced, experienced and indemnified to provide the services required. This will be discussed further in the following chapter.

Engaging a specialist to assist with the health risk assessment and development of risk mitigation strategies is highly recommended.

Integrated health plan (Output)

The primary health planning challenge is to produce an integrated health plan.

The integrated health plan needs to be integrated into all other event plans and coordinate the individuals plans of all organizations involved in health service provision, including:

♦ First Aid

♦ Ambulance, Emergency Medical Service & patient transport providers

♦ On-site medical teams

♦ Local hospital & health service contacts

♦ Regional hospitals and major trauma / tertiary referral centres

♦ Retrieval service providers

♦ Other providers, including welfare, psychological support, pharmacy

It is important to consider the impact health and general event planning on all aspects of event management, especially where common documents are produced. For example:

♦ Maps – all event service providers should work from the same (ideally gridded map)

♦ Contact list – a single list should be issue to avoid duplication, errors and confusion

♦ Emergency plans – all service providers should agree on common emergency plans

♦ Mass Casualty plans – clear safety guidelines and response processes to minimize additional casualties from secondary devices

♦ Forensic scene protection guidelines

The site layout for an event is critically important to health service provision. Some of the biggest problems and barriers to service provision result from a lack of consultation and understanding of the health impact of planning decisions made. Consider the following two event maps, Figure 9.1 and Figure 9.2:

Event layout – Bad

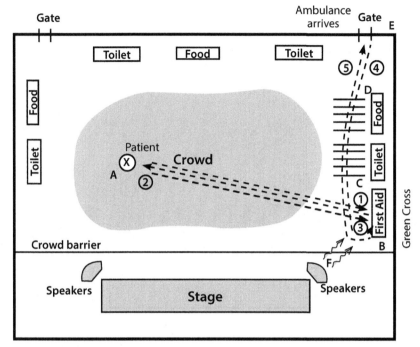

Movements

1 Bystander walks through crowd to First Aid post

2 Response team walks through crowd to patient

3 Assessment in crowd then retrieve patient through crowd to First Aid post

4 Ambulance crew walk through crowd pushing stretcher and carrying equipment

5 Ambulance crew transport patient on stretcher through crowd (minimal patient privacy)

Issues

A Patient location requires response through crowd across stage front (dense crowd)

B First Aid post location
   * Far from exit
   * Deep into crowd
   * Beside toilet (smell, hygiene)
   * No escape from crowd crush

C Toilet queues
   * Block exit from venue
   * May obstruct First Aid post access

D Food queues
   * Block exit from venue

E Venue access gate
   * Far from First Aid post

F Speakers
   * Unacceptable noise level in First Aid post

Figure 9.1: Sample of a bad map

Event layout – Better

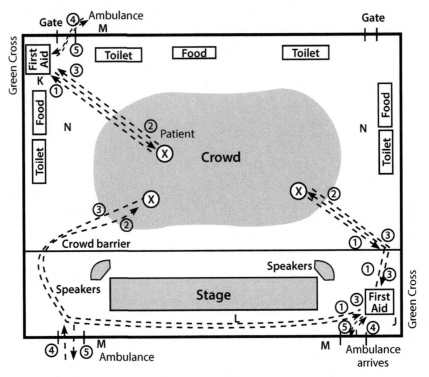

Movements

1 Bystander request for First Aid assistance

2 Response team takes shortest path through crowd to patient

3 Assessment in crowd then retrieve patient through crowd to closest access point to non-crowded area

4 Ambulance crew drive to closest access gate and movce to First Aid post or agreed meeting point

5 Ambulance crew exit venue with minimal need to navigate crowd

Issues

J First Aid post location
   * Out of crowd
   * Behind speakers
   * Close to venue access gate
   * Back stage for cross-stage access (L)

K First Aid post location
   * Additional First Aid post
   * Back of crowd
   * Near venue access gate

L Back of house
   * Cross-stage access

M Ambulance access
   * Venue access gates close to First Aid posts

N Avoid food and toilet queues

Figure 9.2: Sample of an improved map

Example 9.4: Event map comparison for health

In Event Map 1, first aid posts have been located beside the stage co-located with food outlets and toilets. This was in response to requests for power, lighting, running water and close access to toilets (for patient use).

Logistically, grouping power and water outlets together makes sense. Consider then the implications for patients and staff located immediately beside speakers, toilets and potentially crowded by queues for food and toilets. First aid posts and medical centres need access to dedicated toilets for unwell patients, disabled access, ease of cleaning and staff use.

Note also that the exit is at the far end of the fenced-off event precinct, requiring patients to be moved through the crowd (away from the exit) to the first aid post, and then back through the crowd to the exit should an ambulance transport offsite be required.

When plotting a site layout, consult with health and emergency services early and then before finalizing the layout and map, to ensure that:

♦ Crush points are predicted and avoided

♦ Adequate environmental shelter is available

♦ Sufficient sanitation facilities are provided (especially at multi-day and residential events), and

♦ Emergency vehicles access to, from and around the site is maintained at all times.

Communication plan

Responsibility for communications must be clearly assigned, and agreed to, for all phases of an event or incident, including but not limited to:

♦ Health promotion messages before and during events

♦ Liaison with emergency services

♦ Critical incident management

♦ Preparation and distribution of reports, maintain confidentiality in accordance with local legislation

♦ Media management (good and bad)

10 Crowded Health

Introduction

Crowds carry real health risks. By definition, crowds bring large numbers of people in to close proximity and confined spaces. The risk of injury is real, due to accident, crush or malice and the medical risk of disease transmission and demographic-specific presentations must also be considered.

Selecting health service providers is a key early decision. Consulting with local ambulance and health services to build relationships and to seek advice on local providers, legislative requirements and existing health system capacity is time well spent. It is critical that the provider(s) chosen have the skills, resources and experience to service the event and predictable escalation.

Pre-hospital health service provision is a niche industry and is variably regulated. The accumulation of clinical, command and logistical experience takes many years and is a truly heuristic process. A tiered service delivery model, discussed further below, should be adopted with centralized call-taking and management of resources.

Finalizing the size, scope and cost of this model can be a time-consuming and stressful process. This will be informed by the health risk assessment, with mitigation strategies according to ALARP principles, although high consequence outcomes (long tail risks) like cardiac arrest and major trauma will require additional resources.

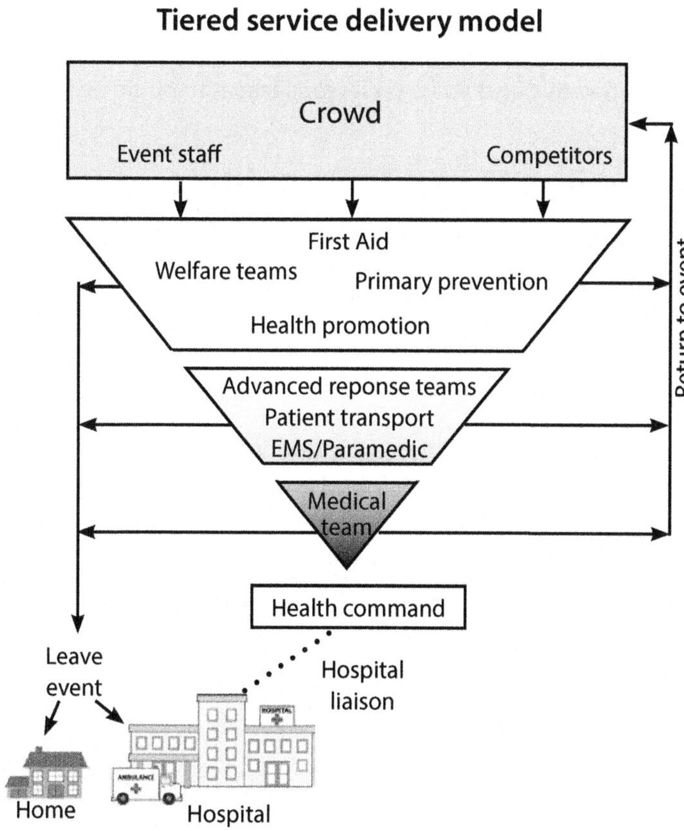

Tiered service delivery model

Figure 10.1: Tiered service delivery model

As events increase in size and complexity, so too will the health service delivery model. Systems and resources must be scalable and plans in place to respond to surges in presentations for medical assistance, including trauma in mass casualty incidents. As more providers are engaged and the health workforce increases, a clear command structure and manageable span of control must be maintained. The Health Command Team will coordinate all health resources and be the direct liaison point with event management. Experienced Health Commanders are experts in factor analysis and making decisions under uncertainty. The Health Command Team will also have direct communication with local ambulance, hospitals and health services.

It is often useful to consider health planning in the traditional risk and emergency management phases of Prevention, Preparedness, Response and Recovery (PPRR). While much of the focus will often be on the response phase, investing in prevention builds resilience, preparedness ensures scalability and capacity to manage surge and recovery operations are restorative and focus on lessons learnt to be carried forward.

Selecting health service providers

Health service is chosen deliberately to include all aspects of first aid, ambulance/EMS (Emergency Medical Services), medical, logistics and health command. Who actually provides the clinical intervention is not as important as the service being provided, although legislation will often dictate and restrict clinical practice in many jurisdictions.

The pre-hospital environment is traditionally the domain of first aid providers and ambulance services. Having a wide variety to choose from can be a blessing and a curse. It is important to check that the organization(s) you choose have the resources and experiences necessary to manage the event. Insist upon being provided with a health risk assessment, operational plans and proof of medical indemnity insurance.

Medical teams, comprising doctors and nurses with out-of-hospital clinical experience, can be extremely useful when engaged strategically within the whole-of-event integrated health plan. These medical teams need to be networked with the local hospitals and health services, to clearly understand each other's capability and scope of practice.

The decision to 'in house' medical services with existing staff or groups of paid or volunteer clinicians should be made with caution. While this may appear to be convenient, flexible and a potential cost saving, established organizations bring experience, culture, systems, equipment, networks and teamwork that cannot be easily or quickly replicated. First aid, ambulance and medical providers should not be tasked with additional responsibilities for the event such as security and fire safety.

Where multiple organizations are engaged, it is essential that a single management structure is agreed upon. Patient flow and clinical responsibility must also be clearly agreed to in advance, especially where there is an overlap in clinical scope of practice between organizations.

Health service provision

Strategic health planning within the PPRR timeline facilitates clear task delegation, increases event health & safety and minimizes the impact of the event on local communities, ambulance and health services.

Prevention

◆ Early liaison with local ambulance, hospitals and health services to:
  ◊ Advise of upcoming event and possible impact
  ◊ Understand local services available
  ◊ Guide planning to augment and avoid overwhelming existing capacity.

◆ Minimizing presentations for medical assistance through health education messaging to increase individuals' resilience and awareness of health risks.

◆ Consultation with public health experts to minimize the impact of:
  ◊ Environmental exposure, through:
    ▪ Provision of adequate shelter from heat / cold / rain
    ▪ Provision of adequate (preferably free) potable water
    ▪ On-site health promotion (e.g. free sunscreen).
  ◊ Infectious diseases through:
    ▪ Tracking and notification processes
    ▪ Implementation of clear sanitation requirements
    ▪ Agreement on response processes.

◆ Consultative approach to event site layout to:
  ◊ Ensure first aid posts and medical centres are:
    ▪ Sufficient in number and appropriately located
    ▪ Fit for purpose, with power, water and toilets as required
    ▪ Clearly identifiable within the event precinct and on maps.
  ◊ Provide clear and protected access and egress for emergency vehicles, to, from and within the event precinct, including pre-determined aircraft landing sites for high risk and rural/remote locations.
  ◊ Avoid convergence areas at high risk of crowd crush

◊ Balance security and health needs, including length of queuing for security screening, drug testing.

◊ Minimize the traffic impact of the event on local communities, especially at the start and end of events where transport movements are highest.

Preparedness (Planning)

◆ Early conduct of health risk assessment.

◆ Security and health plan together to:

◊ Identify and mitigate safety risks

◊ Understand and manage crowd dynamics.

◆ Service delivery model meets needs identified in health risk assessment, including:

◊ Sufficient staff, equipment, facilities and transport vehicles

◊ Appropriate on-site clinical scope of practice to match likely presentations

◊ Capacity to manage long-tail risks.

◆ Engagement with ambulance and health services to facilitate local augmentation of resources as required.

◆ Development of integrated health plan, including single health command structure.

Response (Event Day)

The response phase is where the tools and techniques come together. Staff experienced working in the out-of-hospital environment should be equipped with proven equipment to match their scope of practice.

A dynamic IPO model works well to represent health service operations, with the addition of an output feedback loop feeding the input cycle. This cycling model ensures that Health Commanders and event management maintain situational awareness to guide dynamic resource management while adapting to changes in demand.

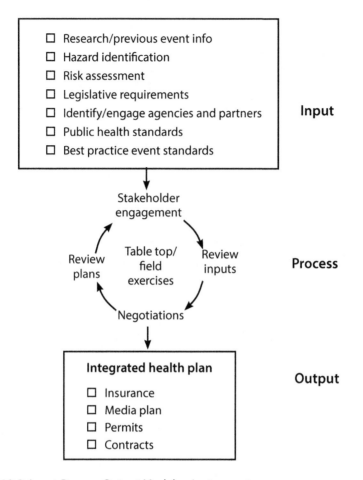

Figure 10.2: Input-Process-Output Model: prior to event

Numerous clinical and operational processes will occur simultaneous and continuously throughout the event or incident, including:

♦ Receiving, processing & documenting requests for assistance

♦ Monitoring event activities, presentation rates & crowd status

♦ Clinical assessment, management & disposition

♦ Response to critical incidents

♦ Providing regular updates to command team & event management

♦ Logistical support activities

Patient flow, demonstrating the dynamic nature of clinical care and the range of outcomes is represented in an IPO model in Figure 10.3.

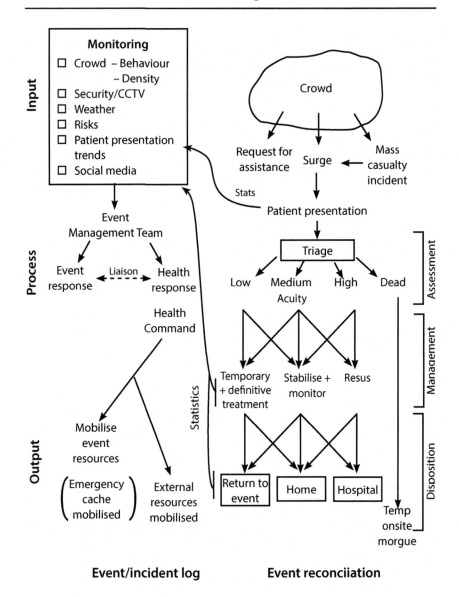

Figure 10.3: Input-Process-Output Model : during the event

Scalability

Scalability is a critical element in emergency response. Ambulance and hospital services are inherently responsive to changes in demand and often run near or at capacity. Experienced clinicians are therefore accustomed to compensating for increased demand, however resources are often limited within the confines of an event boundary. Crowds may complicate gaining access to augment or resupply resources.

Event planning must consider and plan for the need and identify any barriers to rapid changes in levels of services and supplies. Operational systems and management structures should be inherently capable of being scaled up or down as required. Practical considerations include:

♦ Considering methods for increasing staffing levels, including the recruitment of other medical professionals on-site and use of other staff as resource multipliers in response to emergencies.

♦ Reserving overflow patient treatment areas within the event precinct.

♦ Ensuring that all first aid and medical treatment centres are readily accessible.

♦ Maintaining an on-site reserve cache of emergency supplies and equipment.

♦ Developing and exercising a mass casualty plan before event day.

Recovery (Reconciliation)

Paying careful attention to the recovery phase is important to understanding the impact of the event on the local community, ambulance and health services.

A frank review and interpretation of the total number of presentations for medical assistance is an important part of the event / incident debrief and will inform the planning for future similar events.

Health Command

Health Command is an operational role and is importantly distinct from making specific treatment decisions about individual patients: this should remain the responsibility of the treating clinician.

It takes years for a Health Commander to develop the clinical, operational, logistical and management skills and experience required to coordinate and lead a complex functional agency of health workers. While many of the command skills are generic and can be applied in most situations, local knowledge and professional networks are highly beneficial.

Making decisions under pressure and with limited information is pathognomonic in critical care and emergency management leadership. Regular surveillance and routine trend monitoring provide regular updates to guide situational awareness, with additional information provided by exception. A command log should be maintained by all Health Commanders, detailing times and content of key conversations had and decisions made.

Health Command decisions can literally be 'life and death' situations. Some triggers are predictable and should be considered and ideally exercised prior to the event. The impact of decisions on patrons, crowds, staff, the event and local communities must be considered.

It is important that Health Commanders be supported by a well-trained support team and functional systems that can be escalated, avoiding the need to change practice mid-operation. Tasks should be delegated with clear and achievable outcomes and in accordance with good crew resource management (CRM) principles.

The Health Commander should ensure that a 'hot debrief' is conducted after any critical incident, focusing on urgent operational issues and staff welfare. Reports should be written contemporaneously to optimize accuracy of recall.

G Glossary

Some of the technical terms used this textbook that are not common in the events sector. They originate from the military, decision analysis and probability studies. The creation of a theoretical framework for event and crowd management is a rapidly growing field. Hence we have included a short glossary to assist the reader. The explanations are written from the perspective of event management.

Assessment Matrix: also called a *Decision Criteria Matrix* is a table that allows decisions to be made by comparing the situation in the real world with set criteria in tablature form. From this table the supervisor or staff can make a decision and take action to minimise risk.

Bayesian Analysis: a method to make a decision when there are sudden changes. It starts with estimating the probability that an incident will occur, for example a change such as an increase of the density of a crowd. Bayesian Analysis is a simple formula that re-evaluates the probability that the incident will occur now there are new conditions. In real life it is used all the time without people consciously knowing they are using it (an example is crossing the road).

Complexity: is a state with a large number of parts or components and a corresponding large number of relationships between the components. A crowd is an example of a complex entity.

Confirmation bias: when assessing risk we tend to notice evidence that confirms our view point. It is not that we ignore the contrary evidence, we just don't see it or understand its relevance.

Domain: a field of expertise characterised by its own terminology, competency and history.

Emergent behaviour: a crowd is a complex entity and with a result there is behaviour that emerges from the interactions of the crowds. This cannot be simply the sum of individual actions. Complex situations are highly heuristic as they cannot be predicted by linear models. Such behaviour is called 'emergent' as it is far more than a linear sum.

Emergent: a characteristic of a crowd that could not be predicted by looking at each member of the crowd and adding it up. It 'emerges' from the collection of individuals. The complexity of a crowd enables emergent behaviour.

Framework or model: The metaphor of a framework is used to describe the underlying processes. These support the detailed techniques described in the four sections. They have been extracted by comparing each of these four domains. Hence they are descriptive. A model is the next step once the framework has been described. A model is a version of the subject (crowd management) that is small enough to be used to test ideas and developments.

Heuristic : 'rules of thumb' – meaning rules that can only be worked out on the job. It can only come through experience.

Hill climbing: a model of decision making when only partial facts are known. *Local solutions* are found and adjusted with each change. It is analogous to climbing a hill. As the climber moves they readjust their direction according to what they see around them.

Kalman Filter: when two observations are made on the same state, the filter combines the two observations to give more accurate statement. Used in robotics for judgements that are probabilistic. It combines the data and the level or accuracy of the observations to create a more accurate judgement.

Level of confidence (or confidence level): a term used in decision science to estimate the accuracy of any measurement or decision. There are mathematical formulas for this, however in a heuristic model the level of confidence can be given as a probability of being right. When a group of individuals measure a variable or factor, such as flow, the measures will be slightly different.

Linear: when a cause and effect are directly related such that an increase in the size of the cause gives a corresponding increase in the size of the effect. This is common in a stable crowd. For example the more exits there are from a site, the easier it is for the crowds to exit.

Long tail events: rare events or incidents that are unpredictable due to the complexity of the situation, lack of complete knowledge or the impression of causality although it is random. Another name for this is 'Black Swan' (Taleb, 2007). In part, due to these events being unexpected, they can be catastrophic.

OODA (Decision Science): (Observe-Orient-Decide-Act) developed by John Boyd, who modelled the swift interactions occurring between parties or agents in a dynamic environment. See Chapter 5 for a more detailed explanation.

Optimisation: finding the absolute best solution to a problem. *Local optimisation* concerns finding a solution that is the best available at the time. It is important to us as there may be a degree of uncertainty with the measurements and whether the three factors reported by one individual are purely local (just what that person can see). However the nature of time critical decisions means that, although the information will not be perfect, a decision still has to be made.

Silent evidence: when reviewing historical evidence of risk it is well to remember that much information does not survive. An event without any incidents may be random and not due to active risk management. But as nothing went wrong we are not to know.

Situational Awareness: being alive (aware of) to the surroundings, conscious of the changes occurring around as well as the ability to make decisions quickly by assessing and locally optimising the choices. The opposite is to completely focus on one aspect, such as the immediate danger, and subconsciously block out the surroundings, this is also called *tunnel vision*.

State: used in the context of *the state of the crowd*. A state can be described by a list of variables, e.g. happy crowd, flowing freely, moderate climate, informed of drug issues. If these are measurable they can be expressed as a matrix. A change of state is the change of these variables. This can result in emergent behaviour, i.e. difficult to predict exactly, of the crowd. An example is a change in the weather such as lightning for an outdoor festival.

Time-critical decision: a vital term to use for this framework as the decision must be taken very quickly with limited information. In such a situation complete information is impossible and information is limited to observable factors . It is linked to our duty of care – **not taking action is high risk**. The longer the gap between the critical situation and the action to minimise the risk, the higher the consequence of the risk. Time to decide and act on observable information is too often lost in discussion on risk management.

Triggers: Small actions or changes that indicate a looming major change and therefore set off a plan of action.

Variables and factors: *Variables,* in our case, are aspects of the crowd and the surrounding environment that vary with different situations. There are many and it is a very general term. *Factors* are variables in the framework that contribute a risk or problem and its solution.

R References

Abbot, J.L. & Geddie, M.W. (2000) Event venue management: minimizing liability through effective crowd management techniques. *Event Management*, **6**(4):259-270

Adang, O. (2010) *Initiation and escalation of collective violence: a comparative observational study of protest and football events, Preventing Crowd Violence*. Criminal Justice Press.

Agar, M. (1999) Complexity Theory: An exploration and overview based on John Holland's Work, *Field Methods*, **11**(2):99-120.

Allen, J., O'Toole, W., Harris, R. & McDonnell, I. (2011). *Festival and Special Events Management*. Hoboken: John Wiley & Sons.

Australian Disaster Resilience, Handbook Collection, Communities Responding to Disasters: Planning for Spontaneous Volunteers, Handbook 12, Commonwealth of Australia 2017 https://knowledge.aidr.org.au/resources/handbook-12-communities-responding-to-disasters-planning-for-spontaneous-volunteers

Australia-New Zealand Counter-Terrorism Committee (2017), *ANZCTC Active Armed Offender Guidelines for Crowded Places*, Commonwealth of Australia

Berlonghi, A. E. (1995). Understanding and planning for different spectator crowds. *Safety Science*, **18**: 239-47

Bohr, N. (1948). On the notions of causality and complementarity, *Dialectica*, **2**(3-4): 312-319.

Challenger, W., Clegg, W. & Robinson, A. (2009). *Understanding Crowd Behaviours: Guidance and lessons identified*. UK Cabinet Office, 11-13.

Collyer, S. & Warren, C. M. (2009). Project management approaches for dynamic environments. *International Journal of Project Management*, **27**, 355-364.

Festinger, L., Pepitone, A. & Newcomb, T. (1952). Some consequences of de-individuation in a group. *Journal of Abnormal and Social Psychology*, **47**, 382.

Fruin, J. J. (1971). *Pedestrian Planning and Design*, New York : Metropolitan Association of Urban Designers and Environmental Planners.

Fruin, J. J. (1993).*The Causes and Prevention of Crowd Disasters*. First International Conference on Engineering for Crowd Safety, London, England, March. Available at: https://pdfs.semanticscholar.org/467d/5 d641b43f4f7eaca3703e0b2390c60a685b2.pdf

Gell-Mann M. (1994). *The Quark and the Jaguar: Adventures in the Simple and the Complex*, Little Brown and Company, London.

Getz, D., & Page, S. (2016). *Event Studies: Theory, research and policy for planned events*. Routledge.

Goldstein, Jeffrey (March 1999). Emergence as a Construct: History and Issues. *Emergence*, **1** (1): 49–72. https://www.tandfonline.com/doi/abs/10.1207/s15327000em0101_4.

Green Guide (2018) *Guide to Safety at Sports Grounds*, Sports Grounds Safety Authority UK

Health and Safety Executive (HSE). (1999). *The Event Safety Guide: a guide to health, safety and welfare at music and similar events (2nd ed.)*, UK.

Hogg, M. & Vaughan, G. (2008). *Social Psychology*, Harlow, UK: Pearson Education.

Hutton, A., Ranse, J., & Brendan, M. (2018). Developing public health initiatives through understanding motivations of the audience at mass-gathering events. *Prehospital and Disaster Medicine*, **33**(2), 191–196.

International Electrotechnical Commission, International Standard, ISO/IEC 31010:2009, 1st ed., 2009.

Katz, D. & Kahn, R. L. (1966). *The Psychology of Organizations*. New York: HR Folks International.

Kaufmann, S. (1993), *The Origins of Order*, New York: Oxford University Press.

Le Bon, G. (1895). *The Crowd: A study of the popular mind*.

Lightfoot, T. J. & Milne, G., (2003) Modelling emergent crowd behaviour. The Australian Conference on Artificial Life (ACAL). 159-169.

Moussaïd M. & Nelson J. (2014) *Simple Heuristics and the Modelling of Crowd Behaviours*, Center for Adaptive Behavior and Cognition, Max Planck Institute.

Moussaïd, M., Perozo, N., Garnier, S., Helbing, D. & Theraulaz, G. (2010). The walking behaviour of pedestrian social groups and its impact on crowd dynamics. *PloS one*, **5**, e10047.

Murphy, R. F. (1971). *The Dialectics of Social Life: Alarms And Excursions In Anthropological Theory*. New York: Basic Books

NACTSO, N. C. (2017). *Crowded Places Guidance*. UK: Crown.

NERAG (2015) *National Emergency Risk Assessment Guidelines*, Australian Institute for Disaster Resilience, https://knowledge. aidr.org.au/resources/handbook-10-national-emergency-risk -assessment-guidelines/

O'Toole W., (2011). *Events Feasibility and Development: From Strategy to Operations*. Butterworth-Heinemann. Oxford, United Kingdom

O'Toole, W. J. (2018). *Safe and Healthy Crowded Places Handbook*, Australian Institute for Disaster Resilience (AIDR), www.aidr.org.au.

Reason, J. (1990). The Contribution of Latent Human Failures to the Breakdown of Complex Systems. *Philosophical Transactions of the Royal Society of London*. Series B, Biological Sciences.

Reicher, S. (2001). The Psychology of Crowd Dynamics. In: *Blackwell Handbook of Social Psychology: Group Processes*. Hoboken, New Jersey: Blackwell Publishers Ltd. 182-208.

Riley, P. (2014). *Left of Bang*. New York: Black Irish Entertainment.

Schweingrüber, D., Wohlstein R. T. (2005) The madding crowd goes to school: myths about crowds in introductory sociology textbooks, *Teaching Sociology Compass*, **33**:136-153.

Silvers, J.R. (2008). *Risk Management for Meetings and Events*. London. Elsevier Butterworth-Heinemann.

Sime, J.D. (1995), Crowd psychology and engineering, *Safety science*, **21** (1): 1-14

Smith, J. (2010) *Bayesian Decision Analysis*, Cambridge UK.

Snowden, D. (2008). Complex adaptive systems at play (Everything is fragmented). *KM World*, **17**(10).

Snowden, D. (2011). *The Cynefin Framework*. Cognitive Edge. Available at: http://cognitive-edge.com/videos/cynefin-framework-introduction/

Snowden, D. J. & Boone, M. E. (2007). A leader's framework for decision making. *Harvard Business Review*, **85**(11): 68-76.

Standards Australia/Standards New Zealand Standard Committee, AS/ NZS ISO 31000:2009, Risk Management-Principles and Guidelines, November 2009.

Still, G. K. (2013). *Introduction to Crowd Science*, CRC Press, Boca Raton.

Stott, C. J., & Reicher, S. D. (1998a). How confict escalates: The inter-group dynamics of collective football crowd 'violence'. *Sociology*, **32**, 353–377.

Stott, C. J., & Reicher, S. D. (1998b). Crowd action as inter-group process: Introducing the police perspective. *European Journal of Social Psychology*, **26**, 509–529.

Surowiecki, J. (2005). *The Wisdom of Crowds*, Anchor.

Taleb, N. (2012). *Antifragile, Things That Gain from Disorder*. New York: Random House.

Taleb, N. (2007). *The Black Swan: the Impact of the Highly Improbable*, London: Penguin.

The Supreme Council for National Security. (2016). UAE Occupational Health and Safety Management System (OHSMS) National Standard 2016 AE/SCNS/NCEMA 6000:2016,. United Arab Emirates.

Thwink.Org. (n.d.). http://www.thwink.org/sustain/glossary/EmergentBehavior.htm.

UNISDR (2017) *United Nations Office for Disaster Risk Reduction Annual Report 2017*, United Nations

Victorian Bushfires Royal Commission final report, (2009), available royalcommission.vic.gov.au/Commission-Reports/Final-Report.html

WA, (2009). *Guidelines for Concerts, Events and Organised Gatherings*, Department of Health, Government of Western Australia.

Wijermans, F.E.H. (2011), '*Understanding crowd behaviour: simulating situated individuals*', PH.D. Thesis, University of Groningen.

Wijermans, F.E.H., Conrado, C., van Steen, M., Martella, C. & Li, J. (2016) A landscape of crowd-management support: an integrative approach, *Safety Science*, 86:142-164.

Williams. T. (2002) *Modelling Complex Projects*, John Wiley, UK.

World Health Organisation. (2015). *Public Health For Mass Gatherings: Key Considerations*. WHO, Genevre.

Zhang, C., Li, H., Wang, X. & Yang, X. (2015) Cross-scene crowd counting via deep convolutional neural networks. Computer Vision and Pattern Recognition (CVPR), IEEE Conference, IEEE, 833-841.

Index

Printed by Printforce, United Kingdom